DISCOURSE STRATEGIES FOR SCIENCE TEACHING AND LEARNING

This engaging and practical volume looks at discourse strategies and how they can be used to facilitate and enhance science teaching and learning within the classroom context, offering a synthesis of research on classroom discourse in science education as well as practical discourse strategies that can be applied to the classroom.

Focusing on the connection between research and practice, this comprehensive guide unpacks and illustrates key concepts on the role of discourse in students' thinking and learning based on empirical analysis of real conversations in a number of science classrooms. Using real-life classroom examples to extend the scope of research into science classroom discourse begun during the 1990s, Kok-Sing Tang offers original discourse strategies as explicit methods of using discourse to engage in meaning-making and work toward a specific instructional goal. This volume covers new and informative topics including how to use discourse to:

- Establish classroom activity and interaction
- Build and assess scientific content knowledge
- Organize and evaluate scientific narrative
- Enact scientific practices
- Coordinate the use of multimodal representations

Building on more than ten years of research on classroom discourse, *Discourse Strategies for Science Teaching and Learning* is an ideal text for science teacher educators, pre-service science teachers, scholars, and researchers.

Kok-Sing Tang is Associate Professor at the School of Education and the Discipline Lead of the STEM Education Research Group at Curtin University, Australia. He received a BA and MSc in Physics from the University of Cambridge, UK, and an MA and PhD in Education from the University of Michigan, USA.

Teaching and Learning in Science Series

Lederman Norman G. Series Editor

For more information about this series, please visit: https://www.routledge.com/Teaching-and-Learning-in-Science-Series/book-series/LEATLSS

DISCOURSE STRATEGIES FOR SCIENCE TEACHING AND LEARNING

Research and Practice

Kok-Sing Tang

Routledge
Taylor & Francis Group

NEW YORK AND LONDON

First published 2021
by Routledge
52 Vanderbilt Avenue, New York, NY 10017

and by Routledge
2 Park Square, Milton Park, Abingdon, Oxon, OX14 4RN

Routledge is an imprint of the Taylor & Francis Group, an informa business

© 2021 Taylor & Francis

Library of Congress Cataloging-in-Publication Data
Names: Tang, Kok-Sing, author.
Title: Discourse strategies for science teaching & learning : research and practice / Kok-Sing Tang.
Description: New York, NY : Routledge, 2021.| Includes bibliographical references and index.
Identifiers: LCCN 2020022123 | ISBN 9780367369811 (hardback) | ISBN 9780367344245 (paperback) | ISBN 9780429352171 (ebook)
Subjects: LCSH: Science–Study and teaching. | Teaching–Methodology. | Language and education. | Science–Language.
Classification: LCC Q181.A2 T36 2021 | DDC 507.1–dc23
LC record available at https://lccn.loc.gov/2020022123

ISBN: 978-0-367-36981-1 (hbk)
ISBN: 978-0-367-34424-5 (pbk)
ISBN: 978-0-429-35217-1 (ebk)

Typeset in Bembo
by River Editorial Ltd, Devon, UK

CONTENTS

FIGURES

TABLES

FOREWORD

In 1990, I brought together my research from the previous ten years in a book called *Talking Science: Language, Learning, and Values*. The book presented a theoretical and practical approach to understanding the function and importance of talk in science classrooms. It was an early attempt to apply the relatively new field of Discourse Analysis, based on functional linguistics, to the study of how teachers taught and students learned in real classrooms. It emphasized that learning science was in large part learning how to talk the specialized language of science. It reframed the notion of science concepts as interconnected webs of scientific and everyday terminology for talking about natural phenomena. With an eye to the potential usefulness of the results of the research for science teachers and teacher educators, it described a large number of specific discourse strategies found in use by newer and more experienced teachers in fairly typical classrooms.

Over the past 30 years since the publication of *Talking Science*, a substantial body of further research, not just in science education but also in mathematics teaching classrooms and in education in other fields, has added appreciably to our knowledge of not just what kinds of talk and discourse strategies are used in education but also when they work better, worse, or are even counterproductive. Kok-Sing Tang's *Discourse Strategies for Science Teaching and Learning* brings together much of this further research, together with his own original studies and contributions, around the key notion of discourse strategies. The concept of discourse strategies is rooted in the notion that people use language to do things. Teachers use language to convey information, relate abstract concepts to concrete examples, help students to formulate evidence-based arguments, organize classroom activities, and connect what is said in words to what is expressed in diagrams, graphs, and mathematical expressions. As well as much else.

How they do this, and which ways of doing the research has shown to be more effective, is the focus of this new book.

These particular and specific ways of using language to do the work of teaching and learning can be described as discourse strategies. However, every teacher knows, and perhaps science teachers more than most, that teaching strategies do not consist only of ways of talking. Science teachers routinely make use of charts, diagrams, graphs, and for many topics also further kinds of specialized symbols and mathematical equations. This was certainly evident 30 years ago, but at that time the tools to analyze visual and mathematical modes of communication had not yet developed to nearly the same degree as tools for analyzing language, text, and talk. In the time since, many researchers including myself have emphasized the importance of an expanded notion of "discourse" that includes all these other modes of making meaning as well. Tang's excellent presentation of the important new developments in this area leads directly to taking discourse strategies in this expanded sense of mobilizing all the "languages of science" for purposes of teaching and learning.

Other aspects of the processes of teaching and learning in science classrooms for which research has greatly expanded our understanding include: talk and learning in small groups, the importance of using language to help students evaluate and reflect on their own learning processes, the additional needs of learners whose first language is not the language of instruction, and the relationship between discourse strategies to help integrate the learning of minutes and hours into the learning of weeks and months. This book pays attention as well to each of these additional dimensions, both in its theoretical discussions and in analyzing numerous concrete examples of classroom events.

Tang's own research and contributions provide those examples and also highlight the importance of a number of key phenomena, as well as offering practical tools and recommendations for teaching based on his research and the work of others. His own research offers particular insights into the use of metalanguage in the classroom. Metalanguage is just the researcher's term for language that talks explicitly about what we are doing and saying. It is language about language, talk about talk, and more generally talk about activity and communication. Tang particularly emphasizes the importance of how teachers talk about what they are doing and why, where each piece of learning activity fits into the bigger picture, and how students can go about enacting scientific practices, stating scientific arguments, and writing about their science work.

When I wrote *Talking Science*, I made the deliberate decision to put the theoretical basis of the research not at the beginning but at the end of the book. I wanted to make the book as accessible as possible to teachers and teacher educators who were not researchers and did not have a specialized background in functional linguistics or social semiotics. This new book does a very good job of introducing the theoretical background partly at the beginning, but also along the way as it is used and as it can be seen more concretely in the specific episodes of classroom examples. I also rewrote the first two chapters of my book several times based on conversations I had with teachers. Tang's writing here does an excellent job making clear and explicit a long list of key discourse strategies that can be put into practice by classroom teachers. At the same time, it presents clear definitions and explanations of the relevant theoretical background, where it is needed.

A pervasive theme of *Discourse Strategies for Science Teaching and Learning* is making what many good teachers already do more explicit. This is critically important for at least two reasons. First, it makes it more readily possible to convey these effective strategies to new teachers, who might otherwise take years to rediscover them, perhaps without even realizing what their strategies actually are. Second, by making our discourse strategies explicit and providing us with a language for talking about them, we become better able to critique them, to test when and where they do or do not accomplish our goals, and to formulate ways in which we can improve them.

For all these reasons, I highly recommend this new book to researchers, teachers, and teacher educators, particularly in science but also across all disciplines because most of these important discourse strategies are equally fundamental to all teaching. Enjoy reading, thinking about, and making use of this book!

Jay Lemke
Professor Emeritus,
City University of New York

ACKNOWLEDGMENTS

This book would not be possible without the science teachers whom I have the privilege to work with in Australia, Singapore, and the United States. Thank you for letting me share your classroom space and for your courage to try new teaching methods with your students.

I am also thankful for all the pre- and in-service teachers and graduate students whom I have had the pleasure to teach over the years. I am inspired by your eagerness to learn and you have taught me how to better connect my research to practice.

I wish to thank my loving wife, Natasha, for her constant support and encouragement, and my daughter, Isabelle, for bringing the joy that sustains my academic journey. Above all else, I would like to thank my Heavenly Father for His unfailing love and everlasting blessings.

The rich classroom excerpts and artifacts in this book would also not have been possible without the support of the following research projects and grants:

- Developing Disciplinary Literacy Pedagogy in the Sciences, Education Research Funding Program, Singapore (OER 48/12 TKS)
- Formalising Disciplinary Literacy Teaching in Primary Science through a Writing-to-Learn Approach, Education Research Funding Program, Singapore (OER 6/14 TKS)
- Designing Pedagogical Supports for Enabling Inquiry Learning through a Learning Community Approach, Education Research Funding Program, Singapore (LSL 1/04 TSC)
- National Center for Learning and Teaching in Nanoscale Science and Engineering (NCLT), National Science Foundation Grant, U.S.A. (#0426328)

Finally, I wish to acknowledge the following journals for publishing previous versions of some portions in this book:

- Chapter 5 is derived in part from an article published in *Science Education* on 04 May 2017, available online: https://onlinelibrary.wiley.com/doi/abs/10.1002/sce.21275.
- Chapter 8 is derived in part from an article published in *Science Education* on 30 January 2014, available online: https://onlinelibrary.wiley.com/doi/10.1002/sce.21099
- Chapter 8 is derived in part from an article published in *International Journal of Science Education* on 28 January 2019, available online: www.tandfonline.com/doi/abs/10.1080/09500693.2019.1672906

1
INTRODUCTION TO CLASSROOM DISCOURSE

In most classrooms, one would often find a lot of talk and action going on inside. It can be a teacher giving a lecture or students responding to the teacher's questions and instructions. It can also be students talking among themselves as they are doing some written work, setting up an experiment, or exploring a computer simulation. Whatever the activities, talk – *together* with gestures and visual aids – is often the primary mechanism through which the teaching and learning of every subject matter takes place. This is true not only in a language classroom, but also in every content area classroom, and the science classroom is no exception.

No doubt most people will agree that classroom talk is critical to effective teaching. However, there is a general tendency to equate talk with just "communication." This has the unfortunate result of implying that there is something existing (e.g., content knowledge, social norms) that needs to be communicated, and talk is merely a vehicle to transmit this knowledge or practice. Consequently, this creates a narrow view particularly among content area teachers that classroom talk and interaction is auxiliary to the "main" knowledge and practice of the discipline they are teaching.

This book is not about communication in that narrow sense of information exchange or helping students to speak or write properly. It is primarily about *discourse* which is central to the ways of thinking, acting, seeing, believing, and feeling that are associated with a cultural community of people, such as scientists. Without discourse, there is no science. Thus, rather than an auxiliary objective, discourse is indispensable to enabling students to gain access to the knowledge and practice of science. In other words, this book is really about science teaching and learning. In particular, this book is about how science teaching and learning is *constructed* through the most fundamental resource that *all science teachers have* in the classroom – discourse strategies.

Discourse strategies are methods that people from all walks of life use to strategically engage in a meaningful interaction and direct it toward a specific goal. They can range from something as simple as opening and closing a conversation to more subtle techniques involved in various activities, such as bargaining or interviewing. When employed effectively, discourse strategies can be used in science classrooms to: (a) structure conversation to engage students in scientific thinking and inquiry, (b) build content knowledge by linking words, images, and other resources used to represent ideas of the natural world, and (c) equip students with the skills to participate in the social practices of science. Mastering these strategies is key to becoming a skillful teacher of science.

Many of the discourse strategies that will be introduced in this book originated from classroom-based research where researchers, including myself, have observed and subsequently refined the techniques used by experienced science teachers. Some researchers acted as silent observers to record the events that transpired in the classroom, while others were more involved in working with teachers to develop and test new teaching methods and materials. Either way, researchers spent a great deal of time analyzing classroom events in great detail in order to understand how the organization and shared knowledge of classroom life are constructed through talk. Through this research process, experienced teachers' tacit knowledge and experiences were codified and transformed into an explicit form of knowledge that is more teachable and measurable. This iterative process between research and practice formed by the partnership between researchers and teachers plays an important part in advancing our knowledge of classroom discourse and discourse strategies in the science classroom.

Drawing on this research-practice nexus, there are two mutually reinforcing objectives in this book. The first objective is to present a synthesis of the research on classroom discourse in science education in a way that is accessible to non-specialists. This objective will focus on various aspects (or patterns) of *discourse* and how these patterns shape classroom activities and science learning. The second objective of this book is to present several practical *discourse strategies* corresponding to each discourse pattern that could be applied by science teachers into their own classroom practices. This iterative relationship between discourse (as research) and discourse strategies (as practice) will form the backbone of this book. This research-practice connection is reflected in most of the chapters as having the first half focusing on a particular classroom discourse pattern and the second half focusing on a corresponding set of discourse strategies.

Discourse Patterns and Discourse Strategies

To concretize what is meant by discourse and discourse strategy, we will begin by examining an actual dialogue that took place between two people in a classroom:

EXCERPT 1.1

1	Ramona	Okay, why don't you try writing it out? Just now you said when it is rolling ... it has kinetic energy right?
2	Jason	Uh-huh
3	Ramona	So you have kinetic energy
4	Jason	Half m v square
5	Ramona	Okay, so kinetic energy is used tooo ... ?
6	Jason	Produce heat
7	Ramona	Okay
8	Jason	Or sound
9	Ramona	Heat energy due tooo ... ?
10	Jason	Friction
11	Ramona	Okay, very goood ... okay, so now, so K E ... so in other words, when I want to put this into equation, it becomes?

More than Words: Patterns of Science Classroom Talk

From this transcript, two people are obviously talking to each other. However, what I am going to show is that talk is more than just speaking between two people using a common language (i.e., English). Based on certain characteristics shown in this transcript, we can tell that Ramona is a teacher and Jason is a student, as well as what is being taught through the conversation. How do we know these if we were not there during the conversation nor watch a video of it? We know these because we can recognize several characteristic *patterns* that are visible in this conversation. Similarly, Ramona and Jason themselves also intuitively recognized these patterns, which explains how they were able to participate sensibly and smoothly in this conversation.

So what are these patterns that can be found in classroom conversation? The first indicative pattern is based on who is saying what and how. In this exchange, Ramona was asking a lot of questions with a clear intention to elicit some kind of anticipated response from Jason. Jason, on the other hand, was obligingly responding to what he thought Ramona, as the teacher, wanted to hear. Some of Jason's responses were also evaluated by Ramona in the form of "okay" and "very good" (lines 5, 7, and 11). This is the classic Initiate-Response-Evaluate (IRE) interaction pattern where a teacher *initiates* a question, a student gives a *response*, and then the teacher *evaluates* the response. Researchers have long ago found and documented this interaction pattern in almost every classroom (e.g., Mehan, 1979; Sinclair & Coulthard, 1975). While a dialogue based on question-and-answer is common in many other situations (e.g., job interview, panel discussion), this kind of IRE interaction is unique and mostly found in classroom talk. It is precisely due to such interactions that define (and are defined by) the roles and identities of the participants (e.g., teacher, student) as well as the normative rules and expectations found in schools and classrooms. This is one particular aspect of classroom talk and it is defined by the pattern of interaction among the classroom participants, or simply *interaction pattern*. Chapter 3 of this book will further elaborate on the IRE and other interaction patterns.

Another visible pattern that can be seen here is based on what the participants were talking about, and this is related to the content matter of the conversation. Most readers can probably identify the conversation was about physics and energy by looking at several key vocabulary such as kinetic energy, heat energy, and friction. However, readers with a good background in physics will further recognize that Ramona and Jason were talking about a particular physics *concept* called conservation of energy (or work-energy theorem at a higher curricular grade level). This recognition goes beyond just identifying the key words like energy, heat, and friction. It also involves examining the connections among these words. Linguists call these connections *semantic relationships*. There are two important semantic relationships in this conversation. The first relationship is jointly made by both Ramona and Jason in this conversation: KINETIC ENERGY – produce – HEAT or SOUND – due to – FRICTION. The second relationship is an implicit one where KINETIC ENERGY, HEAT, and SOUND are all sub-categories of ENERGY. This relationship is not stated anywhere in the conversation but it is assumed to be known to both Ramona and Jason, who had some prior knowledge in secondary school physics. Now, what makes the concept of energy conservation recognizable to many physics readers in this particular conversation is that those two semantic relationships form a characteristic pattern that is repeated over and over again in many other physics classrooms. It is this pattern that determines whether Jason is correct in his understanding of physics. This is the second aspect of classroom talk and it is defined by the pattern of semantic relationships among the uttered words, or simply *thematic pattern* (Lemke, 1990). See Chapter 4 for more details on semantic relationships and thematic patterns.

There are several other patterns that are characteristic to classroom discourse, but for now, I will briefly mention a *multimodal pattern*. Although spoken language is a key mode of communication here, this conversation would not be possible without the coordinating use of other semiotic (or meaning-making) modes. While this may not be obvious from the transcript, it is only because of the limitation of transcription in foregrounding written words in the expense of other modes. However, from the video that generated this transcript, it is visible that the participants' gazes played an important role in the dialogue. When Jason was responding to Ramona's questions, he was frequently looking at Ramona and his groupmates for some kind of cues to confirm whether he was saying the right thing. At the same time, some of his groupmates (there were four other students sitting at Jason's table) were looking at their laptops which were showing some websites related to the physics concept of energy conservation. Another non-verbal mode that is important here is the use of a mathematical equation to coordinate the conversation. This was seen in Ramona's instructions to "try writing it out" (line 1) and "put this into equation" (line 11), as well as Jason's response to equate kinetic energy as "half m v square" (line 4). The use of these semiotic modes were not incidental; rather they were coordinated with a characteristic pattern. This multimodal coordination pattern complements the interaction pattern and thematic pattern we saw earlier to aid the participants' meaning-making in science. See Chapters 7 and 8 for more details on multimodal patterns.

What Is Discourse?

These examples of discourse patterns illustrate that classroom talk is not just a string of words that is communicated between two or more people, but it involves much more than that. For this reason, the term *discourse* is more appropriate than *talk* in capturing the complex social activity occurring in the classroom, which we have seen three particular *patterns* of. First, discourse always involves social norms, expectations, and identities that are discursively fulfilled through the moment-by-moment *interactions* made by the participants. Second, discourse always involves a particular way of interpreting and talking about our experience and knowledge, which is manifested as the *thematic content* constructed through the use of our words and actions. Third, discourse is always *multimodal* as it involves a range of semiotic modes to coordinate and complement the meaning-making process.

Gee (2011) defines discourse as the combination of language with other socio-cultural practices such as actions, values, beliefs, attitude, and identities within a specific social community. Discourses are deeply embedded in our membership and participation in various communities, and learned through our habitual ways of interacting with people in those communities. They are also manifested in characteristic patterns in the way we speak, write, think, act, and use various tools. A classroom is a kind of social community (meshed within a larger community involving schools and other socio-political entities). As such, classroom (as a collective entity) has historically developed a unique discourse that governs how participants interact with one another, what counts as disciplinary knowledge, and what values and perspectives are preferred over alternative discourses. Central to the theme of this book, the unique discourse in science classrooms can be analyzed and subsequently managed with a more deliberate attempt to enhance science students' learning.

To do so in this book, I conceptualize and define discourse as a social pattern in the use of language that shapes and is shaped by the way we think, act, and make meanings. There are three important aspects in this definition. First, conceptualizing discourse as a *pattern* helps us to see discourse not as a fuzzy concept that is difficult to identify (like "culture"), but as a social phenomenon that can be analyzed and made visible by identifying general patterns from a transcript or video. Identifying these patterns is essentially the key to discourse analysis as a methodological approach. Specifically for this book, I will identify and discuss five different patterns in classroom discourse, of which we have seen three examples earlier: interaction pattern, thematic pattern, and multimodal pattern. The other two discourse patterns are narrative pattern and genre pattern. Each pattern will form the focus of every major chapter (see conceptual framework at the end of this chapter).

The second aspect of the definition, embedded in the phrase "shapes and is shaped by," captures the mutually constitutive relationship between discourse as a global pattern and discourse as an occurring instance. In the earlier example, think about why Ramona was the one asking questions and evaluating the responses. A simple answer is of course she was the teacher. Teachers have always been acting this way and the students have been conditioned to expect this social norm since early childhood. This is the part where discourse (as global pattern) shapes the way Ramona and Jason acted at a particular instance (i.e., asking and answering questions). However, the reverse is also true. What made Ramona a teacher at every discursive moment is also the fact that she maintained, with the cooperation of her students, the kind of interactions that are characteristic of a typical teacher-student relationship in their culture. But imagine now what would happen as every teacher starts to get their students to ask questions while they answer them? This will disturb and eventually change the interaction pattern in the classroom, and *consequently* the roles, expectations, and identities of a teacher and student. This is how discourse can be shaped by recurring instances that are happening over a period of time.

The third aspect of the definition is related to thinking, action, and meaning-making. How is discourse related to the way we think, act, and make meanings? The theoretical basis of this relationship is informed by the integral connection between language and thought. Vygotsky (1986) first made the claim that thought is not merely expressed in words; it comes into existence with them, and thus all higher mental functions are mediated by language. Similarly, Halliday (1993a) argues that language does not simply reflect patterns that are already "out there" as nature but instead imposes the patterns we see in nature by construing a categorical universe of things and relations, which then shape our perception of nature. In other words, the use of language in a discourse is not just a form of expression that "conveys" thoughts and ideas, but it is essentially the ingredients from which the meanings of our thoughts and actions are made of. In Chapter 2, this perspective of discourse and language based on several sociocultural theories will be further elaborated.

What Are Discourse Strategies?

If this is the understanding of discourse, what about *discourse strategies*? Following Gumperz's (1982) pioneering work in sociolinguistics, discourse strategies generally refer to methods that people employ to understand each other in a conversation in order to achieve a particular goal. These methods can be unconsciously employed without much reflection (implicit) or deliberately planned and enacted with a clear target in mind (explicit). In the conversation between Ramona and Jason, Ramona's questioning is clearly a discourse strategy used to direct the conversation toward a specific outcome. The questioning can take the form of asking a direct question (e.g., lines 1 and 11) or inviting the student to fill in the sentence by dragging the last word with a raising intonation (e.g., "used tooo … ?" in line 5, "due tooo … ?" in line 9). These are common strategies used by many teachers. For Ramona, it is likely that these strategies were second nature to her as she used them without much thought and deliberation. When these strategies become second nature as part of a teacher's (or student's) routine repertoire in participating in classroom discourse, I call them an *implicit* form of discourse strategies.

A key argument I will be making throughout this book is that we need to develop more *explicit* forms of discourse strategies for teachers and students to consciously work toward a deliberate target. For example, questioning can be both an implicit or explicit discourse strategy depending on how it is used. When a teacher is asking a lot of questions without much thought to their purpose and how to direct students' responses toward a particular objective, this would be considered an implicit discourse strategy. On the other hand, the use of dialogic questioning, which involves a deliberate, planned, and thoughtful way of framing questions so as to invite diverse ideas and perspectives, can be considered an explicit discourse strategy. (See Chapter 3 for the nature of such explicit discourse strategies in questioning.)

Another way to distinguish between implicit and explicit discourse strategies is to judge whether the discourse strategies were deliberately used or mentioned by the teachers or students. As an illustration, take a look at this excerpt and see if you notice anything different in the discourse:

EXCERPT 1.2

1	John	Explain using the kinetic model of matter, why gas, or air, sorry, air, blow into a balloon inflates it. Notice, it's again an explain question. Now, yup, sorry, you were saying?
2	Wayne	Pro
3	John	So now, since you say pro … What's my P then?
4	Kim	Due to kinematic model of matter
5	John	What's my outcome?
6	Rani	Balloon inflated
7	John	So, or as, or so, the balloon is inflated … So what is the difficult part?
8	Chloe	Reasoning
9	John	Reasoning

In this exchange, we find the usual question-and-answer interaction directed by the teacher, John. However, you may also notice something uncommon about some of the language used by the participants, such as pro, P, outcome, and reasoning. What was unusual in Excerpt 1.2 is that John and his students appeared to have developed a shared language in using these words, and they were using that language to direct and organize the conversation in ways that would not be possible without the use of this shared language. The shared language in this case is called PRO, or Premise-Reasoning-Outcome. It is a form of metalanguage that has been deliberately taught to the teachers through a research project, who subsequently went on to teach all their students (see Tang, 2016a). A metalanguage is a language used to describe the language conventions of a discourse community, which in this context is the language of scientific explanation. The use of this metalanguage is a clear example of an explicit discourse strategy, which was not only used by the teacher, but also the students (notice it was a student who initiated the use of PRO in line 2). The PRO discourse strategy and other types of scientific metalanguage will be further elaborated in Chapter 6.

In this book, I will make explicit several exemplary discourse strategies that I and many other researchers have either documented or developed and tested in science classrooms. It is important to point out that the distinction between implicit and explicit discourse strategies is not a categorical one in order to avoid a false dichotomy between them. Instead, the difference between them is more of a continuum and is dependent on the user and context. Nevertheless, this distinction is useful because whenever I gave the example of using questioning, semantic relationships, or multimodal representations as a form of discourse strategy, I often encountered teachers saying that they are already doing this in their classrooms. In this regard, I make this distinction in order that the discourse strategies presented in this book will be more distinguishable from the usual and implicit routines that are commonly practiced in the classroom. After these discourse strategies are presented from Chapters 3 to 8, I will return to discuss the distinction between implicit and explicit discourse strategies again in Chapter 9.

What Is a Classroom?

The term "classroom discourse" has an unfortunate connotation that it is bounded by what is happening in a physical room. As I will argue in the next chapter, a classroom is not defined as a location but as a cultural space where multiple discourses intersect. In addition, classroom discourse has more to do with what the participants are saying or doing that enact or link to a particular discourse, and less to do with the physical arrangement of bodies, tables, and chairs within the four walls of a classroom. Therefore, science classroom discourse also applies in other settings where science learning activities occur, notably laboratory, field trip, museum, aquarium, planetarium, and zoo.

In the 21st century, classrooms have also increasingly become virtual, as supported by the use of digital devices and computer-mediated communication. In this context, classroom interaction is no longer mediated face-to-face through speech, gesture, and body language. But the interaction is still an integral component that constitutes a "classroom," with interaction now being mediated by a wider range of semiotic modes and technologies, such as synchronous and asynchronous videos, discussion boards, and chat rooms. Speech and gesture are still important through audios and videos, but other modes like written language and visuals will also take on a larger role (Kress, 2003). As such, the role of multimodality will become more important in our understanding of classroom discourse in online spaces (Tang & Tan, 2017). As we explore the theoretical ideas and practical examples of classroom discourse in this book, it is valuable that we do not narrow their applications to only teaching and learning in a physical room.

Who Is This Book For?

This book is primarily written for all science teachers and science teacher educators who are exploring how to translate the research in classroom discourse into practical applications that can improve science classroom teaching practices. The language and discourse of science pose significant challenge to every student who is learning science, even when the student's first language is the medium of instruction used in the science classroom. This is because most students are not familiar with the specialized scientific discourse that is very different from the everyday discourse they use in their daily life (Lemke, 1990). Therefore, the discourse strategies presented in this book are applicable to all primary and secondary school students.

For language minority students, there is an additional language barrier as they have to learn a new medium of instruction (e.g., English) in addition to the language of science, in what is effectively a "multilingual science classroom." Some of the discourse strategies introduced in this book are particularly beneficial for language minority students as they make explicit the language conventions in science that are often hidden. Although this book will be useful for teachers working in multilingual science classrooms, it should be noted that the book cannot provide a comprehensive treatment of strategies geared specifically for language learners, such as the use of translanguaging, codeswitching, and various content-language integrated learning (CLIL) approaches (see Lo & Lin, 2019; Lo, Lin, & Cheung, 2018; Williams & Tang, 2020; Wu, Mensah, & Tang, 2018).

Lastly, this book is also useful for researchers and graduate students who are interested in an overview of science classroom discourse. With a focus on research-practice nexus, this book presents a synthesis of up-to-date research on classroom discourse with an emphasis on practical applications that can be used by teachers to enhance science teaching and learning.

Conceptual Framework and Organization of Book

As we saw from Excerpt 1.1, there are several characteristic patterns happening simultaneously in classroom discourse. This book is organized according to these patterns. Apart from the introduction, theoretical overview, and concluding chapter, each chapter focuses on one of these discourse patterns, which are interaction pattern (Chapter 3), thematic pattern (Chapter 4), narrative pattern (Chapter 5), genre pattern (Chapter 6), and multimodal pattern (Chapters 7 and 8). Each pattern is typically associated with one aspect of science classroom teaching and learning (e.g., activity, content, narration, scientific practice, representation). For example, interaction pattern deals with how activity is structured in the classroom, thematic pattern deals with the content matter of science, and so on. Figure 1.1 shows how this book is organized according to these discourse patterns and corresponding aspects of science teaching and learning.

Each chapter is also divided into two parts. The first part focuses on theoretical ideas and introduces key concepts associated with the discourse pattern in that chapter. These concepts are illustrated with concrete examples taken from conversations in science classrooms. The second part of the chapter focuses on pedagogical applications and introduces a number of discourse strategies that are connected to the corresponding discourse pattern. Many of these discourse strategies were observed or developed from several classroom-based research projects I have carried out over the last ten years. The empirical findings from these projects have also been peer-reviewed and published in numerous scientific articles. For this book, I use a number of excerpts selected from these projects to illustrate and discuss the application of discourse strategies in the science classroom. See Appendix for the research context and methodology that produced the classroom excerpts used in this book.

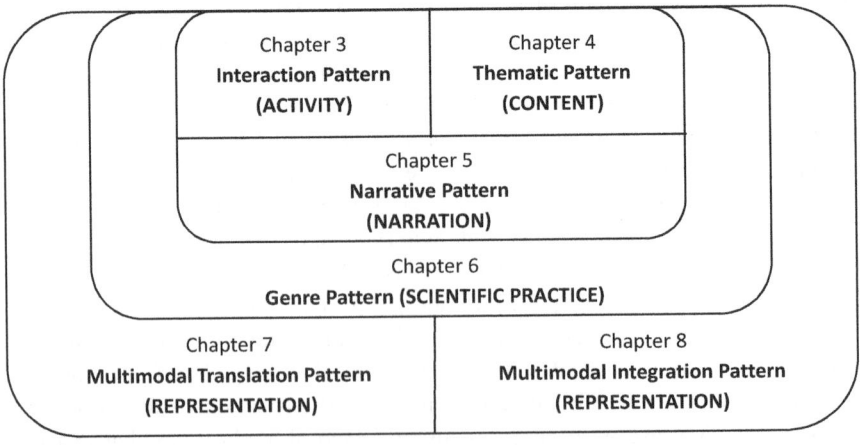

Chapter 3	Chapter 4
Interaction Pattern	**Thematic Pattern**
(ACTIVITY)	**(CONTENT)**

Chapter 5
Narrative Pattern
(NARRATION)

Chapter 6
Genre Pattern (SCIENTIFIC PRACTICE)

Chapter 7	Chapter 8
Multimodal Translation Pattern	**Multimodal Integration Pattern**
(REPRESENTATION)	**(REPRESENTATION)**

FIGURE 1.1 Conceptual framework and organization of book chapters

Chapters 3 to 5 examine the discourse patterns that form the innermost unit of classroom discourse. This organization is inspired by Barnes and Todd (1977), Lemke (1990), and Scott (1998), who posit that science classroom dialogue minimally consists of two patterns – one focusing on the activities engaged by teachers and students and the other focusing on the content in terms of scientific ideas and knowledge. These two aspects of classroom discourse will be examined through the analysis of interaction patterns in Chapter 3 and thematic patterns in Chapter 4. In the conversation between Ramona and Jason in Excerpt 1.1, I have provided examples of the ubiquitous IRE interaction pattern as well as glimpses of a thematic pattern that consists of the semantic relationships among the key words used (e.g., energy, kinetic energy, heat, sound, friction). In Chapters 3 and 4, more examples of these two patterns will be shown, in addition to several discourse strategies that promote classroom interaction and conceptual understanding.

Besides these two aspects of classroom discourse, Scott (1998, p. 58) noted a third aspect consisting of "interventions by which the teacher provides a commentary on the unfolding teaching narrative." This "talking about the scientific narrative" is the focus of Chapter 5, and will be analyzed based on the use of metadiscourse, or meta-talk, to talk about the talk itself. Recent research found that science teachers use a number of metadiscourse to organize, integrate, and signpost the narration of the scientific story that is developed from the interaction and thematic patterns of classroom talk (Tang, 2017). Thus, the judicious use of metadiscourse strategies can help science teachers make the content material more comprehensible, organized, relevant, and engaging to learners.

Chapter 6 examines a broader discourse pattern that shapes and contextualizes the patterns examined in Chapters 3 to 5. This is the pattern of genre, which is characterized by a set of predictable conventions that constitute and enact a particular social practice. There are four major genres in science discourse – experimental report, information report, explanation, and argument (Halliday, 1993a). Although these genres are commonly enacted through science classroom discourse, they are seldom explicitly taught. To do so requires a metalanguage to describe the language used in each of the genres. This chapter will further elaborate on the metalanguage for explanation and argument, and show how they can be used to enable students in explanation construction and argumentation.

Chapters 7 and 8 expand the discussion on discourse patterns and strategies to include non-verbal representations. Although multimodal representations are an integral part of every discourse pattern discussed from Chapters 3 to 6, I temporarily withhold discussing its role from those chapters, and instead dedicate two full chapters for a complete discussion. This is because the same concepts that were discussed from Chapters 3 to 6 (e.g., dialogic interaction, semantic relationship, genre) are applicable for our understanding of multimodal representations. Chapter 7 focuses on the translation of representations from one semiotic mode to another across a teaching sequence, while Chapter 8 focuses on the integration of multiple semiotic modes to make meaning. Specific discourse strategies based on multimodal translation and integration patterns will be introduced to enable teachers and students to coordinate the use of multimodal representations.

2
THEORIES OF CLASSROOM DISCOURSE

The Role of Theories

Paraphrasing from the philosopher Immanuel Kant, Becker (1954, p. 387) once gave a famous quote about the relationship between theory and practice: "Theory without practice is empty; practice without theory is blind." The practice of classroom discourse is one that is occurring in millions of classrooms all over the world. To help us interpret this worldwide phenomenon so as to further improve its practice, we need a theory or theories that can provide us with "a structured set of lenses through which aspects or parts of the world can be observed, studied, or analyzed" (Klette, 2012, p. 4).

In contemporary social science research, there has been a shift away from the positivistic idea or dream that there can ever be an objective "grand theory" that will address all the complexity involved in teaching and learning. Rather, researchers are now increasingly recognizing that theories are intellectual "tool-kits" that we work with to study and understand some phenomena, with an underlying social agenda to change them for the better. Thus, the value of theories should not be casted in terms of what is correct versus wrong, or recent versus old; but rather how useful the theory is in addressing key questions related to the practice of classroom discourse. Some of these questions are:

- How do students learn science through their participation in classroom discourse, and what is the evidence of their learning through the change in their discourse?
- How does the discourse occurring in the classroom relate to an array of social factors, including who we are, what we know and believe, how we feel, and how we affiliate to multiple communities?

- How is scientific knowledge constructed through discourse and how is it related to the broader social and political institutions and communities?
- What is the connection between discourse and other related constructs like knowledge, practice, identity, and meaning-making?
- What is the role of verbal language and other non-verbal semiotic modes in classroom discourse?
- Why are there ideological conflicts occurring in science classroom discourse and how do they arise?

Guided by these questions, what are some of the theories or intellectual toolkits that we will need in this book? In this chapter, I outline a sociocultural perspective to provide a general background and set of lenses to interpret what is going on in the science classroom. Then, I elaborate more specific theories of discourse and language in order to situate and anchor several key constructs that we will use throughout this book.

Sociocultural Perspective

I will first give an overview of a sociocultural perspective or paradigm that underpins the theories of discourse and language that inform this book. According to Anderson (2007) in the *First Handbook of Research on Science Education* (Abell & Lederman, 2007), sociocultural perspective is one of three broad research traditions in science education. The other two are conceptual change and critical perspectives. With its intellectual roots in cultural-historical psychology, sociolinguistics, and anthropology of science, a sociocultural perspective views learning and knowing as an enculturation into the social practices of a discourse community, with science being one of many discourse communities within which people participate in their life (Anderson, 2007).

Why do we need a sociocultural perspective? The reason is not simply because a classroom is made up of a few dozen people talking to one another in a social setting nor are we only interested in collaborative learning. This narrow focus on group learning is just the smallest scale of what we mean by "social" (Lemke, 2001). There are two main reasons why a sociocultural perspective is crucial. First, learning is fundamentally a social activity; even when there is only one person learning alone by reading a book or solving a mathematics problem. Although the learner is not interacting with anyone, he or she is interacting with the language, semiotic tools, and voices from a historical- and cultural-specific discourse community (Bakhtin, 1986; Vygotsky, 1986). Second, all human knowledge, activity, and cognition are embedded and contextualized within larger socio-political institutions and networks, consisting of schools, school districts, government agencies, corporations, research centers, and universities (Brown, Collins, & Duguid, 1989; Latour, 1987; Lave & Wenger, 1991). I will further elaborate these reasons through the seminal works of Lev Vygotsky, Mikhail Bakhtin, and Bruno Latour.

Vygotsky's Semiotic Mediation and Internalization

Lev Vygotsky is a learning theorist whom most teachers have heard about from their initial teacher education but probably misinterpreted or undervalued some of his key ideas. One of his most important ideas is the inseparable relationship between "thought and language," which forms the title of his most influential book. Vygotsky is often quoted for the phrase "all higher mental functions are mediated by the use of signs." What Vygotsky meant by higher mental functions are not a Bloom's taxonomy type of higher-order thinking skills that many educators have come to associate. Rather, higher mental functions are what distinguish human thought processes from all other animals as well as infants and pre-language toddlers. Thus, human thought is not a mysterious thing with an autonomous existence in our mind, but it is formulated through signs; and a system of signs is essentially what makes a language. As Vygotsky (1986, p. 218) puts it plainly, "thought is not merely expressed in words; it comes into existence with them."

Vygotsky also investigated the origin of our inner thoughts from an ontogenetic development point of view. Unlike some of his contemporaries (e.g., Piaget) who believed that cognitive development first occurs on an individual plane before it extends outward to the social plane (e.g., talking to others), Vygotsky argued the opposite. Based on his clinical studies of young children, Vygotsky saw that children's speech development first occurs on a social plane as they observe and interact with adults and older children. They then start to develop egocentric speech (talking aloud to oneself) as a toddler. To Vygotsky, egocentric speech represents a critical intermediary stage where children retell their experiences and play with the language that was used in their social interaction with others. Egocentric speech also functions as the cultural tool to guide the child's thinking and problem solving. Gradually, egocentric speech fades and becomes "internalized" into a silent inner speech as they grow older. In its mature form, inner speech does not resemble social speech. It is characterized by an abbreviation, compression, and agglutination of words, such that a single word is a "concentrated clot of sense" that "is much more heavily laden with sense than it is in external speech" (Vygotsky, 1987, p. 278). These words then become the signs that mediate our higher mental functions as a kind of mental schema. Through this process of internalization, this is how thought develops from a social origin.

Another frequently quoted idea from Vygotsky is the development of "scientific concepts." Vygotsky's use of this term is unfortunate as it confuses with science teachers' common understanding of scientific concepts as defined in the science curriculum (Tang, 2011b). Vygotsky, who is not a scientist or science educator, was more concerned with the role of educational institutions and formal learning in the development of "non-spontaneous" concepts in comparison to spontaneous concepts that children gain from their everyday life. Vygotsky's idea highlights the primary role of systematic instruction and apprenticeship in the development of children thinking, from everyday concepts to systematic concepts. It also highlights that systematic concepts are acquired from a system of knowledge communicated during instruction. In Vygotsky's (1986, pp. 172–173) words, the "very notion of scientific concept implies a certain position in relation to other concepts, i.e., a place within a system of concepts." We will build on this idea of "system of concepts" in Chapter 4 to understand what exactly a scientific concept is.

Bakhtin's Intertextuality and Heteroglossia

While Vygotsky was working on the social origin of the human mind, another Russian theorist, Mikhail Bakhtin, was working on a social theory of language and literary. For Bakhtin, the basic unit of the social lies in the utterance, which is not just a string of words spoken by an independent individual, but a continuous chain within a social *dialogue*. This applies even when there is only one person reading or responding to a text. Any utterance is a retrospective response to preceding utterances and is oriented to future anticipated utterances, thus forming "a link in the chain of speech communication" (Bakhtin, 1986, p. 89). Thus, no utterance can exist in isolation as its meaning only exists in a continual dialogue within a historicity of texts. This central idea was later generalized into a principle called *intertextuality* – the juxtaposition and linkage of texts (Kristeva, 1980; Lemke, 1992). Its basic notion is that a text is never an isolated piece of work, but it always explicitly or implicitly incorporates, assimilates, echoes, challenges, parodies, or responds to prior texts as well as anticipates future production of texts.

Intertextuality comes in two forms – manifest and constitutive intertextuality (Fairclough, 1992). Manifest intertextuality is a direct and overt appropriation of specific texts within a text. This typically includes the use of referencing, quotation, and citation for written texts. For oral speech, metadiscourse markers, such as "according to," "she said," or "yesterday we discussed" are often used in manifest intertextuality (see Chapter 5 for a full discussion on metadiscourse). Constitutive intertextuality is a more indirect and subtle appropriation of broader social conventions, genres, styles, and activity types that are characteristic of distinctive institutions or discourse communities. For instance, to write a laboratory report, a novice will typically use a "template" or study past reports in order to appropriate the structure and style of a laboratory report.

Besides the reciprocal nature of dialogue, Bakhtin (1981) also introduced the notion of *heteroglossia*, which literally means different-tongues. Heteroglossia describes the presence of multiple varieties, or social languages, within a single national language like Russian. These varieties exist due to the social stratification of language use into a variety of ideological belief systems that Bakhtin called *voices*. These voices represent "specific points of view on the world, forms for conceptualizing the world in words, and specific worldviews" (1981, pp. 291–292) of various social groups that reflect stratification according to class, profession, discipline, ethnicity, religion, age group, community affiliation, social movement, and political association.

Due to the intertextual nature of dialogue, no utterance is completely unique and ideologically-neutral. This is because people borrow and adapt others' utterances and voices in constructing their own utterances, and in so doing, appropriate and transform the embedded discourse and ideological stance in their utterances. Every utterance is thus "double-voiced" (Bakhtin, 1981) by blending the voices of others into one's own voice. As a result, heteroglossia is a common phenomenon that can be seen in every text and it reflects not only the ideological belief systems of specific social groups, but also of their sociological relations to one another in terms of their mutual and shifting alliance, connection, contradiction, or conflict. Bakhtin's ideas are central to the discourse patterns that will be discussed in Chapters 3 and 5.

Latour's Actor-Network Theory

Another theorist useful for this book is Bruno Latour, who is instrumental in developing what is now called Actor-Network Theory (ANT). Contradictory to its name, ANT is not a theory but an ontology that describes the reality of human knowledge and sociological structures. ANT originated from the field of science and technology studies (STS) when several ethnographers studied the anthropological life of scientists in the laboratory and fieldwork (e.g., Latour, 1987; Latour & Woolgar, 1979). Their findings revealed a social constructionist view of science that involves a huge amount of literacy tasks undertaken by scientists through the use of *inscriptions*. For instance, Latour and Woolgar (1979) traced the written publications of scientists back to the material resources and actions taken in the research laboratory; from lab mice, chemical substance, apparatus, and machines, to scribbled notes, codified labels, data tables, computer displays, graphs, written drafts, to ultimately a polished academic paper reviewed and accepted by the larger scientific community. Every stage of this *translation* (e.g., from colored samples to a coded table, from scattered points to a continuous line graph) produced a series of inscriptions that selectively and systematically transformed some features of a material substance or data into parts of a larger set of evidence and arguments. This is essentially how evidence are formed through translations to support a scientific claim. There is no mysterious "scientific method" that determines a set of universal rules to be discovered. In Latour's ontology, it is simply the translation of material and semiotic actions through a network of humans and non-humans (e.g., inscriptions, machines).

This network assemblage of humans and non-humans (called "allies") to form a scientific claim extends beyond what is happening inside a laboratory to the larger scientific community. To make one's claim accepted in the community will involve a huge assemblage of allies in terms of convincing peer reviewers, generating support from replication studies, getting citations, and winning continual research funding. Over time, as the network of allies becomes larger, the claim becomes stronger and more convincing. Eventually, it becomes accepted as a *scientific fact* when nobody questions the translation processes and assemblage of allies involved in producing the claim. When this happens, the translation and assemblage will be "blackboxed" until somebody questions them in light of new evidence. But to open this black box will require the assemblage of yet another network of allies by doing almost the same thing; that is, generating more inscriptions and translating the work in another laboratory to published papers, in order to engage in a "trial of strength" with the opposing network. Latour's insight here is crucial to our understanding of scientific practice and multimodal translation pattern to be discussed in Chapters 6 and 7.

Latour's anthropological work on STS was later expanded into a general ontology that describes any social institution or community (Latour, 2005). An interesting insight is the agency Latour ascribes to non-humans. In terms of the propensity to act within a network, a non-human actor has as much influence as a human actor. This radical idea is useful to help us think about the reality of a network that spreads beyond a particular locale and short time-frame. As an individual cannot act simultaneously in multiple places and exist over a human lifespan, it is through non-human actors, most notably as inscriptions that extend the reach of a network over time and space. For instance, the Next Generation Science Standards (NGSS) is a document that was published by the National Academy of Sciences – a powerful socio-political organization in the United States (NGSS Lead States, 2013). It is a document that was produced by an assemblage of politicians, bureaucrats, scientists, educators, as well as other inscriptions (e.g., previous versions, old frameworks). Once NGSS was produced as an inscription, it was circulated widely and subsequently translated to other influential documents, including policy papers, state standards, district curriculum, school syllabus, and lesson plans. This translation and circulation of inscriptions is central to a theory of how discourses are created, legitimized, maintained, and challenged, which I will be developing next.

Theories of Discourse and Social Practice

In Chapter 1, I have defined discourse as "a social pattern in the use of language that shapes and is shaped by the way we think, act, and make meanings" and used a short conversation between a teacher and student to illustrate this definition. In this section, I elaborate further the theoretical basis behind this definition.

Characteristic of Discourse

Discourse is traditionally viewed as simply language in use during communication, focusing on a stretch of language (e.g., sentences, utterances) that are put together in talking or writing. But this view ignores how language is used in social context. The meaning of our communication is intrinsically situated according to a combination of factors, including who we are and the communities we are affiliated to, who we are addressing and their affiliated communities, the language we use and the communities that use them, and what has been or will be said on the topic (Bakhtin, 1981; Vygotsky, 1986). For this reason, the definition of discourse must incorporate these broader social dimensions as well. Gee (2011) coins the term "Discourse" (with an uppercase "D") to encompass larger social affiliations and cultural practices and distinguishes it from "discourse" (lowercase "d") as a stretch of language. In this book, I will not make this upper- and lowercase distinction. I reserve the term *discourse* to include the general features of language in relation to its broader social dimensions and use *text* to refer to a specific instance of language use.

Discourse is the combination of language with the actions, values, beliefs, attitude, and identities within a specific social community. As such, it is also related to broad socio-historical constitutions of knowledge, culture, and power relations (Foucault, 1972; Lemke, 1995). Any discourse is a product of social institutions and networks that construct what we perceive as a social entity (e.g., literacy, physics, schooling) and position people as various social subjects (e.g., blue-collar workers, scientists, students, school administrators). As such, a discourse is also "a construction of some aspect of reality from a particular point of view, particular angle, in terms of particular interests" (New London Group, 1996, p. 25). This "point of view" is the *ideological stance* of the discourse. In a classroom where there are multiple discourses from different people interacting with one another, there is always a diversity of ideological stances reflecting the "various domains of life and experience associated with different voices, positions, and interests."

Gee (2011) further distinguishes a primary discourse as the initial discourse acquired early in life (mostly through our closest community) from secondary discourses that are acquired later in life through our socialization within institutions in the wider communities, such as schools and workplaces. With this distinction, Gee defines literacy as "the mastery of a secondary discourse" (p. 176). As there are always several secondary discourses, literacy is also always plural (i.e., literacies). Thus, literacy is not a set of universal skills for reading and writing text, but it is always connected to a specific form of language within a specific social community. This view of literacy has generated an area of research that examines the specific ways of talking, reading, writing, doing, and thinking in a particular discipline (Moje, 2007). Such *disciplinary literacy* recognizes that all disciplines have different specialized language practices that students need to master in order to gain access to the discourse of that discipline.

Discourse, Text, and Social Practice

Although a discourse is a generalization of some social phenomena, we must bear in mind that it does not exist in the abstract but is manifested as recurring and recognizable patterns in the way we speak, write, think, act, and use various tools. A theory of discourse thus needs to account for this mutually constitutive relationship between discourse as a general "macro" phenomenon and a text as a "micro" manifestation and instantiation of language use. That is, how does discourse create a text, and conversely, how do texts create a discourse? Fairclough (1992) introduces a useful three-level view of discourse consisting of text, discursive practice, and social practice. A text is defined as a product and instance of language use. It is not limited to written words but also includes its spoken and multimodal forms (more on that later). Social practice refers to the broader systems of knowledge, power, and ideology we have discussed earlier. What is critical in Fairclough's idea is an intermediate level called *discursive practice* that mediates text and social practice.

Discursive practice is the process of how texts are produced, distributed, and consumed. Latour's ideas of translation and circulation in an actor-network are useful to understand how texts (as inscriptions) move around multiple discourse communities (e.g., homes, schools, universities, government offices, research laboratories, teacher associations) and connect these communities across time and space. Take the example of NGSS that was provided earlier. NGSS is a text produced by a network of scientists, educators, and bureaucrats with heterogeneous roles. Each of these roles is itself a product and construction of another network with unique social practices that we can identify and call, for simplicity, the discourse of science (for scientists), schooling (for educators), and politics (for bureaucrats). Thus, we can say that NGSS is a text created by and through these discourses.

As NGSS circulates to various communities, through the process of translation in a different actor-network, it leads to the production of other texts, which further get circulated to more networks, and the process continues. Eventually this process makes its way to science classrooms through syllabi, textbooks, instructional materials, and teachers' notes. They then influence how science is being talked or written about in the classroom, including among many things: (a) what counts as legitimate knowledge that must be learned, (b) the conduct of activities that mirrors scientific inquiry and practices, and (c) the design of formative and summative assessments. This is how the discourses of science, schooling, and politics (to name a few) shape the production and interpretation of texts in the classroom.

This process does not stop here as the texts produced in science classrooms are part of the discursive practice that feeds back into the larger discourses. An example is the collation of students' texts for assessment purposes, such as standardized tests. These student texts are typically translated into numbers and statistics for various assessment reports at a school, district, state, or national level. Some of them are also collated from a multitude of countries and translated into international comparison reports such as TIMSS and PISA. As these texts are produced, they have the power to change the network, leading to a revision or creation of other influential texts, such as the next version of NGSS or equivalent policy papers in other countries. As this network of text production, distribution, and consumption changes over time, it also changes the discourses that are associated with it. Thus, from Latour's ANT perspective, a discourse is not a larger social structure that exists independently from texts, but it is a complex and dynamically changing network of texts and people. For instance, I have analyzed in a previous study how the discourse of "scientific practice" as enacted in a classroom is constructed as an actor-network of literacy events performed by teachers and students (see Tang, 2019c).

The translation of text across discourses also involves people frequently moving in and out of multiple communities. In one community, they may "internalize" the language and "appropriate" the voices from the discourse of that community (Bakhtin, 1986; Vygotsky, 1986), and subsequently reproduce those language and voices in another community. A good example can be seen in teacher education where most preservice science teachers typically receive training in subject specialization (e.g., physics, biology, general science) and educational studies as well as undergo internship through school placements. Through this process, they encounter and participate in multiple discourses like science, education, and school administration. These discourses are not always in harmony and they often create conflicts for the preservice teachers' knowledge and identity formation (Larsson, Airey, Danielsson, & Lundqvist, 2018). As the teachers graduate and enter science classrooms, they act as key mediators in introducing the voices of multiple discourses into classroom discourse. In the same way, students also participate in multiple communities other than schools (e.g., home, church, club, online community) and appropriate the discourses associated with them (Tang, 2011a). These processes explain how various discourses shape language use in the science classroom. Consequently, the science classroom is also a cultural space where multiple discourses meet and interact through the voices that are appropriated by and projected through the teachers and students.

Theories of Language and Meaning

A theory of discourse provides a toolkit for us to interpret the connection between language use in the classroom to the larger social practices, institutions, and communities. However, we still need a theory of language that accounts for how people make meanings with the linguistic and semiotic resources they have under different circumstances. This is where I turn to the theories of systemic functional linguistics (SFL) and social semiotics. SFL provides not only a theory of language to explain how meanings (including scientific knowledge) are made, but also the methodological tools that I use to analyze the language of teachers and students in some chapters. Social semiotics is a more general theory that largely develops from SFL (Lemke, 1990) and it expands the scope of language from a previously linguistic-dominant mode to all semiotic modes (e.g., images, gestures, physical objects).

SFL and social semiotics are extensive theories with a vast amount of literature that cannot be sufficiently summarized in this book. For a more comprehensive treatment of these theories as applied in science education, readers can consult the books *Scientific Literacy for Participation: A Systemic Functional Approach to Analysis of School Science Discourses* by Knain (2015) and *Multimodal Teaching and Learning: The Rhetorics of the Science Classroom* by Kress, Jewitt, Ogborn, and Tsatsarelis (2014). For the purpose of this book, I will only cover several key concepts in this chapter, such as semiotic system, metafunction, choice, context, realization, genre, materiality, and affordance.

Language as Semiotic System

Michael Halliday was a linguist widely known for his work in SFL. One of his early insights was to move away from the view of language as a universal syntax (c.f. Chomsky) and recognize that language is a tool for people to make meanings in various social settings. Halliday (1978) describes language as a *semiotic system* that is used by a particular culture to serve its interests and communicative needs. From this functional role, language evolves into a resource with three meaning-making functions (Halliday, 1994). These three functions (or metafunctions) are: (a) ideational – using language to construct and represent our ideas and experiences of the world and our own consciousness, (b) interpersonal – using language to enact our interaction and relationships with others as well as express our stance toward other people and our experiences, and (c) textual – using language to organize text itself by connecting parts of a text into a coherent whole. Any use of language (i.e., text) will always draw on these three metafunctions of language to make ideational, interpersonal, and textual meanings simultaneously under different social contexts. Chapters 4, 6, and 8 will focus on ideational meanings as we explore how science content is built from language. Chapters 3 and 7 will focus on the interactional side of interpersonal meaning, while Chapter 5 will examine the attitudinal side of interpersonal meaning (as evaluative metadiscourse) as well as textual meaning (as organizational metadiscourse).

In Halliday's view, a semiotic system is not just a collection of signs, but a systemic network of interrelated *choices* that people make from the sets of options available to them through the materiality of the signs within the system. Choice is the "mechanism for expressing meaning by creating a contrast between what is chosen and what is not but could have been" (Fontaine, Bartlett, & O'Grady,

2013, p. 3). For instance, when a person observes an event, the ideational function of verbal language allows him to say (or think to himself) something like for example, "It is dark." Every choice of word classifies this event into some categories in relation to other categories from contrasting words (e.g., dark vs. bright, it vs. she, is vs. was). This consequently narrows the meaning about the event from a range of all possible meanings, or "meaning potential," into something more specific. Of course, these three words are insufficient to represent the event and it is still open to multiple interpretations of what those words mean. So more words need to be used. The addition of more words does not actually *add* more meanings to the observation, but rather, through the choices of what were said in relation to what were not said, the representation of what the observation *could be* become narrower and more precise.

The above example might suggest that in observing an event, we form a thought first before expressing it through words. But the process is actually two-way where the language in our disposal shapes what we can perceive or think about the event. Going back to the example of the "dark" event, we can only think about the event in terms of all the possible categorical relations that could be made within the language that we use. This reinforces what I had mentioned in Chapter 1 that language is not a vehicle that transmits a pre-existing thought or idea, but rather every thought, including our perception, is shaped by language. As Halliday (1993a) puts it, "language does not simply reflect patterns that are already 'out there' as nature. Rather, it imposes the patterns we see about nature by construing a categorical universe of things and relations, which then shape our perception of nature." This view of language is also consistent with Vygotsky's (1986, p. 218) notion that "thought comes into existence with words."

Meaning and Contexts

Meaning must always be made in context. Another way of saying this is any text is only meaningful when we connect the text to its context (Lemke, 1990). In SFL, text and context are not treated as independent entities, but are mutually constructed and analyzed (Knain, 2015). This relation between text and context is built into SFL theory through the concept of *realization* and it is typically represented visually as a series of co-tangential circles shown in Figure 2.1 (Martin, 2014).

SFL stratifies language use through a number of levels with increasing abstraction. Each higher level is *realized* by a patterning of the lower level. At the lowest level is phonology and graphology, which are the system of sounds and symbols we use to speak and write respectively. Above this level is the system of words or lexicogrammar (lexicon plus grammar). This system of words is realized by a pattern of sounds or symbols from the lower level. For example, in most writing systems that use the Latin alphabet, every word is realized from a combination of 26 letters. The next level is discourse semantics which is the system

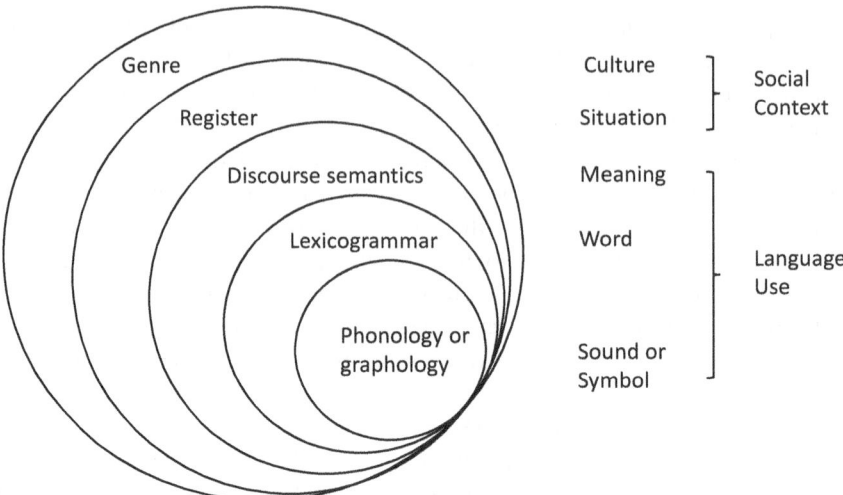

FIGURE 2.1 Levels of realization from text to context

of meaning, and it is realized by a combination of words from the lower level. For example, the meaning of any clause or sentence is realized from a combination of all the words known in that language. I will also provide a more concrete example of how meaning is made with words in Chapter 4. Essentially, meaning-making at the level of discourse semantics is realized through a combination of words that are put together; in other words, a pattern of lexicogrammar. And a pattern of lexicogrammar is realized through a pattern of phonology and graphology. These three levels of phonology/graphology, lexicogrammar, and discourse semantics constitute to what we often associate with language use.

Above the level of discourse semantics is where we need to consider the social context. There are two nested levels of context here which Halliday (1985) calls "context of situation" and "context of culture." The context of situation comprises the external circumstance that influences the choices made by a person in interpreting or producing a text. This includes what is happening when the text is read or produced, who is saying or doing what, and what is the mode of communication? These choices are explained by the notion of *register*, which is the semantic variation of language use that is typically associated with a configuration of situational features (Halliday & Hasan, 1976). For instance, when a teacher is verbally explaining a phenomenon involving forces and movement (the situation), there are often a number of semantic variations in his talk and action (the text production). These could be a frequent use of action verbs, conjunctions, animated gestures, and passive voice. On the other hand, if the teacher is making a joke (another situation), the semantic variations will change to something else. These variations of language use are a higher-order pattern of discourse semantics.

The second level of context extends beyond the immediate situation to the broader context of culture within a community. This is where we associate meanings to the way language is typically used to achieve the goals and purposes of a recurring type of social activity. In SFL, this realization is explained by *genre*, which is a culturally evolved way of doing things with language. Genres are all the repetitive activities that people do in their social life, including dining at home or at a restaurant, buying things from a shop or online, attending a lecture, seminar, concert, or church service. As these genres are quite consistent within a culture, this is how we can seamlessly participate in them by predicting how the activity will unfold and expecting what to say and do, and what not to say and do, at various stages. When we move to a different culture and encounter a slightly or vastly different genre for the same activity, we often have to adjust and sometimes relearn to participate in them.

In the earlier example, I explained how the activity of giving an explanation in the classroom (situation) gives rise to certain semantic variations (register) in the language use. At a higher level of abstraction, this activity itself is often recurring with predictable stages and expectations within the culture of a science classroom. This recurring pattern is how we are able to recognize the situation as a particular type of genre called scientific explanation. We are also able to distinguish this genre against other recognizable genres (e.g., experimental report, scientific argument), which are defined by a different set of patterns associated to a different social activity. The concept of genre and the four most common genres in science will be further elaborated in Chapter 6.

Therefore, in any culture, we can identify a system of genres that characterize the general ways of how language is used to achieve specific goals for specific activities. Each genre is a pattern of register, which in turn is a pattern of discourse semantics, and a pattern of lexicogrammar, and a pattern of phonology for speaking or graphology for writing. These "patterns of patterns" explain how meaning is contextualized at different levels (Lemke, 1984).

Multimodality

Meaning is never made with verbal language alone. In the development of SFL, although Halliday was mainly concerned with the linguistic system of the English language, his central idea that language is a semiotic system is applicable to all semiotic modes. Based on the same concepts of system, choice, and realization, a number of SFL researchers began to broaden Halliday's theory to include other semiotic systems, called *modes*. These modes include images (Kress & van Leeuwen, 1996), music (van Leeuwen, 1998), movement and gesture (Martinec, 2000), and mathematical symbolism (O'Halloran, 2000). The term *multimodality* (multiple modes) thus arises from the study of this multiplicity of modes used in all human meaning-making (Jewitt, 2008).

As in the case of verbal language, every semiotic mode consists of a system of culturally shaped signs (e.g., sounds, shapes, symbols, objects) that are used and molded over time by a particular community into a meaning-making tool. In the history of human civilization, multiple modes have been developed from the use of these signs in different cultures (e.g., writing, image, gesture, music, Braille). All these modes function as meaning-making resources with the same three metafunctions – ideational, interpersonal, and textual. The *materiality* of the sign greatly determines the kind of meaning that can be made from a particular mode. According to the semiotician Peirce (1986), every sign requires a material substance for it to be made and communicated. By shaping and working on the material substance over time in order to meet certain communicative requirements in a community, this is how every mode was developed. For example, speech and music as a mode is built on the materiality of sound, while writing is built on inscriptional devices (e.g., paper, screen). Some modes like gesture and dance rely on the movement of our bodies and hands to make various narrative and symbolic meanings. Braille as a mode relies on the materiality of uneven surfaces (e.g., embossed paper) to create a writing system based on tactile sensory.

The inherent nature of the material determines the specific potentials and constraints of every mode, and this gives rise to the *affordance* of the mode. For instance, the temporal characteristics of human sound both afford and constraint speech as a mode to make meanings in a sequential manner. On the other hand, the spatial characteristics of visual drawings allow meanings to be made and interpreted simultaneously instead of sequentially, thus allowing a unique way of meaning-making compared to speech. As a mode is intrinsically linked to its materiality, the development of the material through technological progress can expand the meaning-making potential of a mode. For instance, in the evolution of writing from the use of stone tablets, to ink and paper, to the digital page on modern devices, each technological improvement brings with it new ways of making meaning that was not available previously.

In terms of how meaning is made in context, the same principle of "patterns of patterns" shown in Figure 2.1 also applies to every semiotic mode to some degree. For instance, in a visual mode, the system of phonology and graphology (by sounds and letters) in a verbal mode is replaced by a system consisting of line, space, and color as its fundamental units. At the level of lexicogrammar, instead of words for verbal mode, this will be replaced by shapes, which are realized by the patterns of lines, spaces, and colors. Thus, just as we can analyze the meaning of a spoken or written text based on the pattern of its words (to be done in Chapter 4), we can also analyze the meaning of a diagram based on an equivalent pattern of its visual elements. Examples of how this principle applies for a visual representation will be shown in Chapter 8.

Beyond the level of discourse semantics, it is possible to extend the same reasoning to register and genre of a particular mode. We can thus imagine multiple sets of concentric circles similar to Figure 2.1 for every single semiotic mode. However, the idea that there is a system of genres to a single semiotic mode is probably only applicable for a narrow range of specialized discourses; for example in classical music, where every genre (e.g., concerto, sonata, fugue, waltz) is realized by a pattern of musical compositions, which are further realized by notes and rests. In science and science classroom discourses, we hardly make meanings with just one mode alone. Verbal language, visual images, and mathematical symbolism have historically co-evolved to function as an integrated whole in constructing a scientific view of the world (Lemke, 2003; O'Halloran, 2000).

Due to the inseparable integration of verbal language with all other modes in most discourses, it is not helpful to regard multiple independent systems of genres for each semiotic mode. Rather, it is more realistic and productive to conceive a multimodal integration of all modes as one generalized system (Tang, 2013a). Thus, a scientific genre, such as an explanation or experimental report, is never realized through verbal language alone as it often involves other non-verbal modes as well. Conversely, non-verbal modes, such as scientific diagrams, graphs, and equations, are hardly used on their own outside the context of a genre and its purpose (Tang, 2019a). As multiple modes are integrated at the level of register and genre, the meaning potential they afford teachers and students to make expand exponentially. This corresponds to Lemke's (1998, p. 92) multiplying meaning principle whereby the "meanings made with each functional resource in each mode can modulate meanings of each kind in each other mode, thus *multiplying* the set of possible meanings that can be made."

Summary

Through the theoretical lenses as described in this chapter, discourses are combinations of language with sociocultural practices that are products of social institutions and networks. Through discourses, we construct our worldview of reality and position people in different social relationships from a particular ideological stance. As such, knowledge, beliefs, attitudes, and identities are always embedded within and enacted through discourses. Discourses arise from our membership and participation in multiple communities, and are maintained by the production, distribution, and consumption of texts in and across communities. As people and texts move in and out of multiple communities, this is how different discourses interact in a social space, such as a classroom. Depending on the discourses' ideological stances, the interaction of multiple discourses may result in harmony or conflict.

Discourses are manifested as characteristic patterns in the way we use language in a specific context. Identifying these characteristic patterns is vital to revealing the underlying discourses and how they influence and shape our actions and meaning-making. It is also the key to the use of discourse analysis as a methodological approach. Informed by SFL and social semiotics, the patterns of language use can be analyzed at different levels according to how a higher level is realized by a patterning of a lower level; that is, genre is a pattern of register, which is a pattern of discourse semantics (meaning), which is itself a pattern of lexicogrammar (words). This "patterns of patterns" (Lemke, 1984) theorization also applies for non-verbal modes where instead of words as comprising sounds and letters, it will be shapes comprising line, space, and color for a visual mode, and bodily movement comprising body parts with pace and space for a gestural mode, and so on. At the same time, language has three metafunctions – ideational (content), interpersonal (participation and stance), and textual (organization). Hence, each of these metafunctions has the potential to manifest as a different discourse pattern.

This theorization of a discourse and its characteristic patterns are the foundation that grounds my definition of discourse as "a social pattern in the use of language that shapes and is shaped by the way we think, act, and make meanings." It also informs this book's conceptual framework of distinguishing five discourse patterns, namely, interaction pattern, thematic pattern, narrative pattern, genre pattern, and multimodal pattern. Each of these patterns and their associated discourse strategies will be further expanded from Chapters 3 to 8.

Finally, I argue that it is only by revealing these discourse patterns that we can gain a conscious awareness and understanding of how discourses work. By doing so, we can then develop more *explicit* forms of discourse strategies for us to consciously and deliberately use to make scientific meanings. This will ultimately allow us to have more control over the patterns with the short-term goals of improving our teaching practices and long-term goals of changing the discourse in the classroom.

3

USING DISCOURSE TO

Establish Classroom Activity and Interaction

Imagine you are having a dinner conversation with a friend and you have something specific to tell her. How would you broach the subject? If you want to get her attention, you might preface your message with something like "Hey, I have something to tell you." If you like to get her intrigued first, you might ask a question like "Do you know what happened today?" Or you might simply cut to the chase and say "I got an award today!" Following the social norm in most cultures, your friend will likely respond with something nice or ask a question, and this will then provide a cue for you to further elaborate your award. However, if your friend is not listening or responding in ways that you expected, you will probably "repair" the conversation by asking "Did you hear what I say?" or "What's wrong?" Regardless of the content of the message you are going to say, there is always *some kind of structure* that frames how the message is being communicated to another party.

All social conversations have such a structure, even for something as simple as communicating a short message like getting an award. Researchers in discourse analysis have studied numerous conversations in everyday context to understand how they are constructed through verbal language, which according to Halliday (1985), has an interpersonal metafunction that enables speakers to interact and negotiate with one another in a conversation. The research has identified that most conversations have a basic exchange unit. A common type of exchange unit is an adjacency pair (Sacks, Schegloff, & Jefferson, 1974) consisting of two turns made by two different people such that the second turn is always dependent on the first. Common examples of adjacency pairs are: question/answer, statement/acknowledgment, offer/accept, and command/compliance. The structure of many conversations is made of a series of these adjacency pairs. Socially speaking, there is no way for any message to be communicated without such a structure.

Classroom talk is a kind of social conversation. As such, it also has a structure consisting of a series of exchange units that most teachers and students follow. However, it has several distinctive conditions that set it apart from other social conversations (e.g., friendly chat, job interview, shopping transaction). One such condition is the need to disseminate knowledge and managing a large group of children or adults. Consequently, classroom discourse has evolved and developed a unique exchange unit called "triadic dialogue" (Lemke, 1990) that consists of three turns instead of an adjacency pair. We have already seen an example in Chapter 1 when Ramona the teacher was asking her students a series of questions through an IRE (Initiate, Response, Evaluate) interaction pattern – teacher initiates a question, a student responds to the question, and the teacher evaluates that response. The IRE pattern is the most common exchange unit used by many teachers in their whole-class conversation, despite the fact that most of them are not even aware of such conversational structure.

Many educators have criticized the IRE pattern as a restrictive structure that limits students' voices and thinking. However, the IRE pattern itself is not intrinsically bad for student learning. The issue is whether teachers are aware of such interaction patterns, and thus have control over them such that they can shift or modify the interaction patterns according to the needs of the students and the transition of instructional activities in a lesson. Unfortunately, science teachers are usually more concerned with the "message" (or content) of the conversation, instead of how the conversation actually takes place through interaction. But as this chapter and the next chapter will show, these two aspects – content and interaction – always occur together simultaneously; just as any content needs an interaction (as a structure) to be communicated, an interaction cannot exist without some kind of content in the communication (Lemke, 1990).

The objective in the first half of this chapter is to outline several interaction patterns and their characteristics, beginning with the basic IRE pattern. This is then followed by a variant of the IRE pattern called IRF (Initiate, Response, Follow-up) as well as a discussion of teacher talk and student dialogue. Once the various interaction patterns are introduced, the second half of the chapter focuses on how teachers can shift their interaction patterns to make their lesson more interactive and dialogic. Specifically, discourse strategies that promote IRF and student dialogue will be introduced and discussed with examples.

Interaction Patterns among Classroom Participants

As elaborated in Chapter 1, classroom discourse is conceptualized in this book as a form of social pattern in the use of language that shapes and is shaped by teaching and learning in the classroom. This book will introduce five different discourse patterns, beginning with interaction pattern as the first and most fundamental of all the patterns. Interaction pattern examines the conversational exchange between two or more people, and raises questions concerning their mutual roles and expectations, such as who is talking, who controls the talk, how do the speakers coordinate their talk, what kind of questions or statements are being made, and so on. In this chapter, we will explore four major interaction patterns commonly found in a classroom, namely IRE, IRF, teacher talk, and student dialogue.

Initiate-Response-Evaluate (IRE)

When a teacher is talking science to the whole class or a group of students, the IRE pattern is a very common way to involve them in the conversation, as seen in the following excerpt:

EXCERPT 3.1

#	Speaker	Utterance	Move
1	Teacher	Nathan, what is the definition of pressure?	I
2	Nathan	Pressure is the amount of force acting per unit area	R
3	Teacher	Okay, pressure is the amount of force acting per unit area. Let's put it into a formula, so our P, which is our pressure, is equal to force over area	E
4	Teacher	Zac, what is the S.I. unit for force?	I
5	Zac	Newton per square meter	R
6	Teacher	Force. Not pressure	E/I
7	Zac	Oh. Newton	R
8	Teacher	Newton, alright. So the unit for force is Newton	E
9	Teacher	Area?	I
10	Zac	Square meter	R
11	Teacher	Square meter or meter square	E

In the above excerpt, there are four consecutive IRE exchange units. Each exchange unit began with the teacher asking a question. This is identified as the *Initiate* move as shown in lines 1, 4 and 9, as well as line 6 when the same question from line 4 was implied. The purpose of an initiate move is to elicit a response from the students. The response could come from a specific student asked during the initiate move, as seen in this example (e.g., Nathan, Zac) or any student volunteer when the question is an open invitation. Finally, the teacher made the third move, which is called the *Evaluate* move as it would give an indication of whether the previous response was considered valid or not. If the response was correct, the evaluation could take the form of a praise, an acknowledgment like "okay" (line 3) or "alright" (line 8), or a repetition of what the student had said (line 11). The evaluate move would close the exchange and the teacher could then move on to the next question and started a new IRE exchange.

If the response was incorrect or not acceptable, as happened in line 5, the teacher would indicate so in the next move. This could take the form of a rejection or hesitation (e.g., no, try again, uh . . .), or repeating the question or part of the question as in line 6 (e.g., the question asks for "force, not pressure"). In most classroom culture, when a student answers a question wrongly, there is often an implicit invitation to try again. This explains why Zac immediately provided another response in line 7 even though the teacher did not ask the question again. Thus, the evaluation made by the teacher in line 6 also served as an initiate move for the next IRE exchange from line 6 to 8. As soon as Zac gave the correct answer, the teacher evaluated it positively in line 8 and closed the IRE exchange for that particular question (i.e., unit for force) and moved to the next question (i.e., unit for area).

From this example, we can identify two important characteristics of an IRE interaction pattern. The first characteristic is the type of question that was initiated by the teacher. In an IRE, the question is usually a close-ended type of question that requires a simple and factual answer, such as the definition of a scientific term (line 1) and the units of measurement (lines 4 and 9). As there is only a narrow range of "correct" responses to a close-ended question, the third move by the teacher has to be somewhat evaluative as we saw in line 6. Thus, an IRE interaction pattern typically involves: a close-ended question to initiate an exchange, a narrow range of responses, and an evaluation that either signals correct or incorrect.

The second notifiable characteristic is the authoritative nature of an IRE interaction pattern. Based on Bakhtin's (1981) notion of voice, a conversation is considered authoritative if there is only one point of view or "voice" and dialogic if there are multiple voices (Mortimer & Scott, 2003). As explained in Chapter 2, a voice represents a specific worldview from a discourse rather than the physical speech from a person. As such, there can be one voice even when two or more people are talking (authoritative), and likewise, one person can be talking or thinking about an idea from multiple voices (dialogic). In an IRE pattern, the teacher only accepts a limited range of answers that are aligned with one particular point of view, while rejecting other views given by the students. In doing so, the teacher maintains full control of the conversation and there is very little room for discussion because alternative points of view are not accepted by the teacher.

In most cases, the first characteristic of an IRE – close-ended question – reinforces its second characteristic – authoritative talk. This is because when the initiate move asks a close-ended question and the response from a student is either right or wrong, the teacher will either accept or reject that response. However, there can be some cases where a close-ended question may lead to a dialogic talk, and conversely, an open-ended question can still lead to an authoritative talk. For example, it is possible that when a student gives a response that is not totally wrong or only partially right, a teacher may still choose to be authoritative and reject the response. Likewise, when a student provides a wrong answer according to the scientific point of view, the teacher may still choose to withhold the evaluation (at least for a little while) and maintain a more dialogic approach to the conversation. In the next interaction pattern, we will see an example of this approach.

Initiate-Response-Follow-up (IRF)

The IRF interaction pattern is very similar to IRE in terms of its triadic structure and the roles of the participants. However, there are some key differences in its characteristics as a result of a small adjustment in the last move from *Evaluate* (E) to *Follow-up* (F). To compare this interaction from the previous IRE pattern, we will examine a class discussion shown in the following excerpt. The question that drove the discussion was: "When you are drinking water with one straw inserted inside the water and another straw outside the water, why is it more difficult to drink compared to just one straw inside the water?" Prior to this discussion, the teacher performed a demonstration of the phenomenon and drew a diagram as shown in Figure 3.1.

Looking at Excerpt 3.2, the first notable difference is the type of question that was asked compared to the previous IRE interaction. The question was first asked in line 1 in conjunction with a demonstration by the teacher to show the difficulty of sucking the water in the "second scenario." (The first scenario involved the teacher drinking with one straw, and it was demonstrated and discussed before this interaction occurred.) Unlike the IRE pattern,

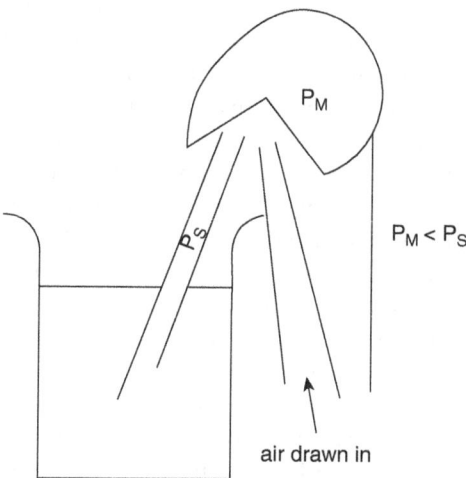

FIGURE 3.1   A diagram showing the phenomenon that was discussed

EXCERPT 3.2

#	Speaker	Utterance	Move
1	Teacher	Okay, so now this the second scenario, where I'm using two straws. So why is it that now that I have two straws, as observed by this motion (*demonstrates sucking the water*), it's very difficult for me to drink the water? So, why is that so? Cindy?	I
2	Cindy	When you're drinking, the straw that's outside is drawing in air	R
3	Teacher	Okay, so air is being drawn in. So when we are drinking, or a better term, when we are sucking, air is being drawn in drawn in from through the straw on the outside. Okay, good, and then?	F
4	Cindy	Thus increasing the pressure in the mouth	R
5	Teacher	So now, here, there is air from here. So she says she's increasing the pressure, thus increasing pressure in mouth. And then?	F
6	Cindy	So now the pressure in mouth is higher than the pressure in straw. So	R
7	Teacher	She's saying the pressure in the mouth, pressure in the mouth (*writes P_M in Figure 3.1*), is higher than the pressure in the straw (*writes P_S in Figure 3.1*). She's saying that P.M., P.M. is actually greater than P.S. (*writes $P_M > P_S$*). And then?	F
8	Cindy	So, difficult to suck	R
9	Teacher	What do you all think of her answer? I'm not saying it's wrong. So, I'm just saying what do you think of her answer when she says that P.M. is greater than P.S.?	F

this question was evidently more complex as it involved a "why" question that required more thinking and elaboration. More importantly, this question was left open-ended for Cindy to answer. The teacher did not break down this complex question into smaller parts as a form of scaffolding to help Cindy answer the question step-by-step. If he had done so, this would turn the interaction into an IRE pattern with a series of short and simple questions.

Corresponding to this type of open-ended question in the initiate (I) move, the first response (R) given by most students is unlikely to address the question completely. Thus, the purpose of the third move is to "follow-up" (F) with the student's partial or incorrect response. For example, Cindy gave a response in line 2 that was accurate but insufficient to explain the phenomenon. Thus, the teacher repeatedly asked "and then?" three times, as seen in lines 3, 5, and 7 in order to get Cindy to *extend* her response. Simultaneously, another follow-up move used by the teacher was *paraphrase*, which drew on the ideas from Cindy and redirected them into an appropriate language. For example, the teacher explicitly quoted Cindy's ideas by saying "she says" (line 5) and "she's saying that" (line 7). He also paraphrased her words from "pressure in the mouth" and "pressure in the straw" to P_M and P_S respectively in line 7. These symbols were then written next to the diagram on the board and used in later conversation.

In the teacher's follow-up move, although there were some evaluative elements such as "okay, good" in line 3, the focus of the move was to keep the conversation going instead of closing it with an evaluation that served as an affirmation or closure. In fact, at one point, the teacher deliberately withheld his evaluation of Cindy's response, even though it was wrong when Cindy said that the pressure in the mouth (P_M) was higher than the pressure in the straw (P_S) in line 6. Instead, he paraphrased Cindy's incorrect response in line 7. Then, in line 9, he used a follow-up move called *reflective toss* to direct the evaluation of that response to the rest of the class instead. Therefore, the follow-up move in the IRF interaction has a different discursive function compared to the evaluation move in the IRE interaction. The various types of follow-up move (e.g., extend, paraphrase, reflective toss) will be further discussed in the second half of this chapter on discourse strategies.

As the function of the follow-up move is to continue the conversation, this creates a continuous chain of I-R-F-R-F pattern (Mortimer & Scott, 2003). In this chain, the meaning of the follow-up moves (F) can only be understood against the overarching question that was first initiated (I). For instance, all the *extend* (e.g., and then?) and *paraphrase* questions we saw in this example only made sense in light of the overarching question they were trying to explain. In most cases, this I-R-F-R-F chain often continues until the teacher feels that the first (initiate) question has been addressed and the class is ready to move on to a new question. By contrast, the IRE pattern typically occurs in short and discrete units that do not form a continuous chain. In the previous example, all the questions asked by the teacher were standalone questions (e.g., definition, S.I. unit) such that the responses (R) and evaluations (E) for those questions occurred within each IRE exchange unit, and they did not spill over to adjacent IRE units. Therefore, comparing between IRE and IRF patterns, we can say that IRF is more dialogic as it allows a longer dialogue, thus giving students more opportunities to voice and consider their points of view. Furthermore, when a student gives an incorrect response that contradicts the scientific point of view, an IRF interaction aims to build on the student's idea and gradually steer it toward the scientific view, instead of brushing it off as wrong in an authoritative IRE interaction.

Teacher Talk

The third interaction pattern besides IRE and IRF is teacher talk, which as the name suggests, is an extended monologue made by the teacher. In most instructional contexts, teacher talk is also called a lecture as the teacher provides information on a topic with no or very little input from the students. Although the teacher does all the talking, teacher talk is technically still an interaction, as defined by a mutual exchange between the teacher and the students with their respective roles and normative behaviors. In this case, there is a clear line between who is doing the talking as opposed to who is (and should be) listening. While there is no verbal exchange, the interaction is largely mediated through non-verbal cues such as gaze, facial expression, and bodily movement. For example, a teacher may stop his lecture if he or she notices that most of the students are not looking at him.

The following excerpt shows an example of a teacher talk, which occurred as a very long utterance at the end of the class discussion that we saw in the previous IRF interaction pattern. The discursive move in this case is not an initiate (I) or evaluate (E), but a statement (S):

EXCERPT 3.3

#	Speaker	Utterance	Move
1	Teacher	It has to be higher, so that the difference is lower. All these are things that you need to consider step by step, especially when you draw your answer, especially like what she writes here, okay, about the air being drawn in part, and then, you know . . . it increase the pressure here, P.S. (P_S), yes, you increase the P.S., but the pressure cannot be bigger than the P.A. (P_A), the P.A. must still be bigger so that it'll cause the water to go in. It's just that in this case, pressure difference decreases, it's more difficult, or it takes more effort to drink the water. Okay, the main thing to take note of atmospheric pressure is just how this works, how to formulate your answer, and then, the most important thing, atmospheric pressure equal to 1 atm or 101,300 Pascal.	S

This teacher talk provided a summary of the earlier discussion. Most of it is propositional statements of the physics content (e.g., pressure) and some paraphrases students' earlier inputs. For instance, if you recall, the term P_S (pressure in the straw) was re-voiced from Cindy's response. There is, however, no question asked in this teacher talk that is directed to anyone to answer. Finally, there are some instances of meta-talk, which directs an audience to certain features of the talk itself (e.g., "all these things you need to consider," "main thing to take note," "most important thing"). Meta-talk or metadiscourse is another discursive resource, which we will explore in more detail in Chapter 5.

Student Dialogue

While IRE, IRF, and teacher talk are dominated by the teacher, the last interaction pattern – student dialogue – is largely mediated among two or more students. There are two approaches that teachers generally use to incorporate student dialogue in their lessons. The first approach is to nominate a student or group of students to present their work to the class, typically after some time has been given for them to prepare the work. The student's presentation to the class often occurs as an extended talk by a student (similar but opposite to teacher talk), and occasionally followed by a few exchanges of question and answer between the presenter and the rest of the class.

The second approach is to divide the class into pairs or groups and have them engage in a discussion. In this approach, the interaction pattern among the students is much more varied and dynamic compared to the more controlled dialogue in an IRE, IRF, or student presentation. Moreover, the interaction moves in a student dialogue typically manifest as adjacency pairs instead of a triadic dialogue. This is because there is usually no central person who controls and facilitates the discussion similar to what a teacher typically does in an IRE or IRF interaction. The common adjacency pairs that occur in a peer-to-peer dialogue are: (a) state/acknowledge and question/answer for giving and demanding information, and (b) offer/accept and command/comply for giving and demanding goods and services (Halliday, 1985).[1] Let us examine the discursive moves and adjacency pairs in the following example when two students were discussing a particular question:

EXCERPT 3.4

#	Speaker	Utterance	Move
1	Shanti	This P.S. (P_S) is bigger, right?	State
2	May	Why?	Question
3	Shanti	Because air is drawn in the straw what . . . so more air, more pressure	Answer
4	May	Does more air mean more pressure?	Question
5	Shanti	You see here right . . . the air moving this way. Bigger pressure equals more force in the water	Answer
6	May	I don't know . . .	Acknowledge
7	May	Should we ask Mr Tan?	Offer
8	Shanti	Okay	Accept
9	May	Can you ask?	Command
10	Shanti	Fine	Comply

In line 1, although Shanti's utterance was framed as a question, if we consider the context of the preceding and subsequent utterances (Bakhtin, 1986), she was not really demanding information. Instead, she was giving information as a suggestion to the discussion question, "Is P_S in the two straws scenario bigger or smaller than the P_S in the one straw scenario?" The "right?" in Shanti's utterance made it sound like a question, but in terms of its function, it was actually used to express her uncertainty toward her suggestion. However, Shanti's statement in line 1 was not acknowledged until later in line 6.

In line 2, May was not certain about Shanti's earlier statement, so she asked a genuine question requesting for why Shanti thought P_S was higher. This was fulfilled by Shanti in the following turn (line 3), which was then followed by another pair of question and answer in lines 4 and 5. Finally, after the series of clarification from line 2 to 5, May remarked that she was still not sure that P_S was higher. May's remark in line 6 served as an acknowledgment and closed the adjacency pair to Shanti's statement first proposed in line 1.

From line 7 onward, the exchange took a different turn. Here, instead of exchanging information, there was an action first suggested in line 7 (i.e., calling the teacher) and eventually realized. Thus, the adjacency pairs were made to facilitate the giving and requesting for a kind of service. These adjacency pairs were offer/accept in lines 7 and 8 and command/comply in lines 9 and 10.

Transition across Interaction Patterns

In this chapter, I have presented classroom interaction pattern as four distinct types: IRE, IRF, teacher talk, and student dialogue. However, it must be noted that this distinction only serves a heuristic purpose of communicating and understanding a complex social phenomenon. In reality, the boundaries of these interaction patterns are much more fluid and overlapping. Therefore, it is more useful to consider these interaction patterns as a continuum instead of discrete categories. On a scale from more teacher control and authoritative to more student involvement and dialogic, the range of the continuum is: teacher talk, IRE, IRF, and student dialogue. With this continuum in mind, we can use it to understand how a teacher can shift their interaction toward more student involvement or more teacher control in order to achieve various instructional purposes within a lesson.

As you might have noticed by now, all the excerpts used for every interaction pattern earlier actually occurred within one physics lesson focusing on the topic of air pressure. Table 3.1 shows the transition of the dominant interaction patterns and their respective duration and purpose within that lesson:

TABLE 3.1 The transition of interaction patterns within a lesson

Duration (in mins)	Dominant Interaction Pattern	Purpose
0.5	Teacher Talk	Present lesson objectives
1.0	IRE (Excerpt 3.1)	Recapitulate key terms and concepts
1.5	Teacher Talk	Set context for discussion question
		Demonstrate phenomena of drinking with one straw and with two straws

(*Continued*)

TABLE 3.1 (Cont).

Duration (in mins)	Dominant Interaction Pattern	Purpose
6.5	Student dialogue	Think-pair-share discussion
10.5	IRF *(Excerpt 3.2)*	Whole-class discussion
4.0	Student dialogue *(Excerpt 3.4)*	Think-pair-share discussion
2.0	IRF (shifting toward IRE at the end)	Whole-class discussion
2.0	Teacher Talk *(Excerpt 3.3)*	Summarize the discussion and learning

As shown in the table, there was a range of interaction patterns used in the lesson and each interaction pattern served a different purpose. Thus, there is nothing inherently bad about any of the interaction patterns, particularly teacher talk or IRE, which is often regarded as limiting student learning. However, when a lesson excessively and predominantly consists of just one interaction pattern, then there is limited variation that can suit the changing requirement of different stages in a lesson. In other words, a skillful classroom teacher should be able to use a wide range of discourse strategies to facilitate these various interaction patterns.

Interaction Discourse Strategies

We have examined the first pattern of *discourse* in a typical science classroom, which involves four types of interaction pattern enacted by the participants. Corresponding to each interaction pattern, there is a wide range of *discourse strategies* that the participants employ consciously and unconsciously to facilitate their interaction. For instance, in an IRE or IRF, a teacher often uses various techniques to call on different students to respond to his or her questions. Many of these discourse strategies are implicit (e.g., unconscious, intuitive) as they are used without much thought and deliberation. Moreover, most teachers pick up these discourse strategies through the "apprenticeship of observation" (Lortie, 1975) after spending many years in K–12 classrooms as students themselves observing their own teachers. This explains why many beginning teachers tend to ask questions in class following a rudimentary form of IRE pattern, even though this pattern was rarely taught in most teacher education programs.

The problem with observation apprenticeship is that many discourse strategies remain *implicit*, and as such are not judiciously used to direct the interaction toward a specific instructional goal. Without a conscious awareness of discourse strategies, a teacher may ask too many short and evaluative questions that do not promote student dialogue and thinking or give too little "wait time" for students to respond. Therefore, this book aims to make several exemplary discourse strategies *explicit* for science teachers to facilitate meaningful interaction with the students or among the students. In particular, because classroom discourse tends to be predominantly authoritative, strategies that are more useful in facilitating dialogic interaction via IRF and student dialogue will be the focus for the rest of this chapter.

The discourse strategies in this chapter are closely related to questioning, which is a topic studied in many teacher education programs. Research in teacher questioning has found that teachers ask hundreds of questions on an average day (Morgan & Saxton, 1991). However, most of these questions are either managerial types of questions or simple recall questions following a typical IRE interaction. A study by Wragg and Brown (2001) found that only 8% of the questions asked by teachers promote student thinking and reasoning. As such, there is a need to ask more dialogic questions to involve students in the thinking process. There are three ways to achieve this objective. The first is to promote dialogic questioning by asking more open-ended questions in an IRF interaction. The second is to follow up the dialogic questioning with a chain of follow-up questions or discursive moves. The third approach is to encourage students to ask collaborative questions to one another through student-student dialogue.

Dialogic Questioning with Open-ended Questions

Asking the right dialogic question is key to promoting deep learning in general as well as enabling scientific inquiry. As we saw from the earlier examples, an open-ended question provides more opportunities for more dialogic interaction and in-depth discussion while a close-ended question tends to result in brief IRE types of responses and affirmations.

What are good open-ended questions in science? Open-ended questions are those that elicit a range of views, complex ideas, or multiple relationships and factors. In science, questions that involve "why" or "how" are usually more open-ended as they require logical reasoning, analysis, and justification. Of course, these questions should be somewhat new to the students such that they do not already have the answer, otherwise the questions will just become simple recall and comprehension. Here are some examples of open-ended questions targeted at various levels of understanding and complexity:

- When 1 cm^3 of sugar is added to 10 cm^3 of water, why is the total volume less than 11 cm^3?
- Why does a raisin sink and float repeatedly when it is dropped in soda water?
- Why does the top side of a pizza (with the cheese) seem hotter than the bottom (crust) even though both sides are the same temperature when the pizza is removed from the oven?
- Why did dinosaurs become extinct?
- Is the potato we eat a root, fruit, or stem of the plant? Why?

In these examples, some students may often give a one-word response (e.g., dissolve, density, heat capacity, asteroid, stem) and think that they have answered the questions. However, the knowledge of a few keywords does not imply a good understanding and thus many students need to be prompted to provide more explanation or justification. For instance, by just providing asteroid collision as the answer for dinosaur extinction, it does not sufficiently explain why only non-avian dinosaurs are wiped out while some plants and animals have survived and evolved till this day. In the example that asked about the classification of potato, although the "scientific answer" is a stem, it is important to discuss other alternatives based on its properties and functions against the criteria of classification between a stem and a root. Thus, these open-ended questions provide ample opportunity for critical thinking and discussion through an IRF or student dialogue.

In addition, many questions asked in K-12 science education often have an acceptable answer according to scientists. However, when there are alternative answers or explanations, even though they may turn out to be wrong, they should be encouraged in order to foster dialogic (multi-voiced) perspectives, at least in the beginning. In the pizza example, although heat capacity is the most likely explanation, thermal conductivity is also possible and the extent of its effect in comparison to heat capacity could be discussed. In the dinosaur extinction example, the asteroid collision is currently the most popular explanation accepted by most scientists and known to the general public. However, within the scientific community, there are alternative explanations (e.g., volcanic eruption, climate change) that are competing with the asteroid explanation. These alternative explanations for both the pizza and dinosaur extinction questions provide useful materials for students to engage in scientific argumentation. Similarly, the classification of potato as stem tuber (along with other interesting examples such as Pluto, jello, and plasma) also provides opportunities for multiple perspectives and debates. Overall, questions that foster scientific explanation and argumentation provide suitable open-ended questions. Further details concerning how scientific theory, claim, and evidence work within the genres of explanation and argument will be discussed in Chapter 6.

Dialogic Questioning with Follow-up Moves

In an IRF interaction, while open-ended questions set the stage for the initial question (I) in a dialogic conversation, appropriate follow-up (F) questions are needed to provide the support for students to build the knowledge to address the initial question. In the first half of this chapter, we saw an example of how a teacher used several follow-up questions (e.g., paraphrase, extend, reflective toss) to support Cindy in developing an explanation for an open-ended question – why it was more difficult to drink water with one straw inside the water and another straw outside the water. As emphasized in the following excerpt, these follow-up questions were chained in a way to continue the conversation with the students, and at the same time, steering it toward the teacher's instructional objective.

EXCERPT 3.5

#	Utterance	Follow-up Move
1	[Opening question]	
2	[Student's response]	
3	"So when we are drinking, or a better term, when we are sucking . . .	Paraphrase
	And then?"	Extend
4	[Student's response]	
5	"So she says she's increasing the pressure, thus increasing pressure in mouth . . .	Paraphrase
	And then?"	Extend
6	[Student's response]	
7	"She's saying that P_M actually greater than P_S.	Paraphrase
	And then?"	Extend
8	[Student's response]	
9	"What do you all think of her answer?"	Reflective Toss

In the following, I will introduce five types of follow-up moves with examples to illustrate how they can be used to support a dialogic I-R-F-R-F chain of discussion (Mortimer & Scott, 2003). These moves are called *extend, probe, paraphrase, reflective toss,* and *constructive challenge.* They are also collectively known as "Socratic Questioning" (Chin, 2007).

Extend. Whenever a student gives an incomplete response to a question, one of the most intuitive follow-up moves is to ask for elaboration by extending the response. This *extend* move continues the initial reasoning given in the previous response and is different from asking for a different response or reasoning. Besides the previous example, here are two more examples of this common follow-up move through prompts like *and then?, what else?, so how?,* and *so what?*:

EXCERPT 3.6

#	Speaker	Utterance	Move
1	Teacher	Anyone brave enough to tell me why is carbon monoxide considered poisonous?	I
2	Jimmy	It will combine with red blood cells	R
3	Teacher	Okay, it will combine with the red blood cells **and then?**	F
4	Jimmy	It reduces their oxygen	R
5	Teacher	Yeah. Alright.	F
6	Jimmy	Preventing from carrying the oxygen.	R

EXCERPT 3.7

#	Speaker	Utterance	Move
1	Teacher	Ben, how is shadow formed?	I
2	Ben	When the . . . something blocks . . .	R
3	Teacher	What is the something?	F
4	Ben	Object	R
5	Teacher	Okay, when I ask you how is shadow formed. You cannot say when something is block by the object. **What else?**	F
6	Ahmad	Blocked by an opaque object	R
7	Teacher	**So how?** My question is how is the shadow formed? So when light is blocked by an object? **So what?**	F

An *extend* move works very well in a scientific explanation that involves a *why* or *how* question. This is because a scientific explanation consists of a chain of logical reasoning that is joined together through cause-and-effect consequence or temporal sequence (Tang, 2016a). For example, in explaining carbon monoxide poisoning, the reduction of oxygen in red blood cells (Excerpt 3.6, line 4) is a consequence of the red blood cells reacting with carbon monoxide molecules. Therefore, when a student only provides an initial cause (e.g., reaction with red blood cells) without completing the next consequence, the *extend* move prompts him or her to move forward in the reasoning process until the explanation given has sufficiently accounted for the phenomenon to be explained (e.g., carbon monoxide poisoning, shadow formation).

Probe. Another common move to follow up on an incomplete response is *probe*. Although *probe* is somewhat similar to *extend*, it works in an opposite direction. While *extend* moves the reasoning forward toward a specific consequence or outcome, *probe* moves the reasoning backward to an underlying cause, evidence, or source of knowledge. Consider the following example:

EXCERPT 3.8

#	Speaker	Utterance	Move
1	Teacher	So what is the reason for a solid not being able to conduct electricity?	I
2	Stella	No free-moving ions	R
3	Teacher	Good. There is no free-moving ions. But you got to explain a little bit more. **Why?** Because there's no free-moving ions, and **why are the ions not moving?**	F
4	Stella	They are bonded together to one another	R
5	Teacher	Very good, so you notice there are two things that you mentioned here. Number one, there is no free mobile ions. And number two is because of strong electrostatic forces of attraction that holds it together, okay?	F

In this explanation, the consequence of "no free-moving ions" (line 2) is that a solid is unable to conduct electricity. If the teacher had asked an *extend* question like, *and then?* or *so what?*, the student would probably give the logical conclusion that is already stated in the question (i.e., unable to conduct electricity). However, by asking a *probe* such as *why?*, this prompts a backward reasoning to identify a more fundamental cause or principle. In this case, the "bond between ions" (line 4) is the cause for not having free-moving ions. It is a general principle that applies to the arrangement of molecules in most solids. Of course, this reasoning process does not end there and further probing questions can be asked concerning why the molecules in solids have strong forces of attraction, and even beyond that until no further question can be asked anymore once the participants have reached the limits of their knowledge. Figure 3.2 shows the difference between *extend* and *probe* as a follow-up in relation to the logical sequences of the explanations given in the examples.

Besides probing for the underlying cause or principle that provides the basis of an explanation, *probe* can also be used to discuss the evidence, assumption, or source of knowledge that inform the students' responses. Prompts in this case typically include: *How do you know that?*, *Where do you get that from?*, and *What is the evidence?* For example, by probing continuously for the cause of how a shadow is formed, the discussion will eventually arrive at the fact that light travels in a straight line. This statement is one of the key assumptions in the ray model of light (a 17th century invention in geometrical optics). As such, "light travels in a straight line" cannot be explained any further because there is no other underlying cause. However, it can be supported or refuted by empirical evidence, and the appropriate probing question to ask in this case could be, "What evidence or experiment suggests that light travels in a straight line?" For more discussion concerning the nature of scientific explanation, theory, argument, and evidence, refer to Chapter 6.

Paraphrase. Paraphrase is a powerful follow-up move that can complement a teacher's repertoire of discourse strategies. A paraphrase is more than just

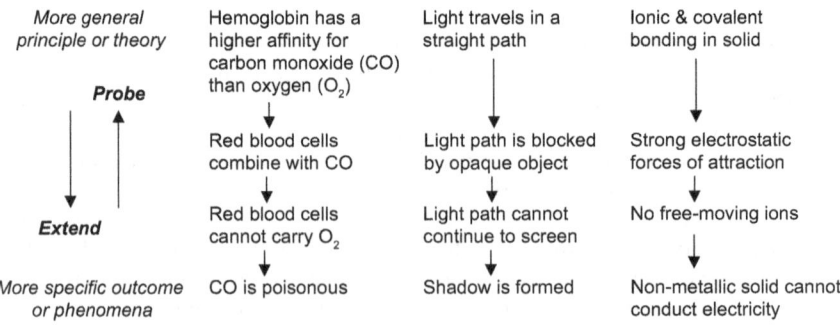

FIGURE 3.2 Logical sequences of explanations as prompted by probe or extend

a repetition of students' ideas, it also incorporates two functions that are subtle to most participants. First, paraphrase selects and summarizes pertinent portions or aspects of students' voices in a way that is productive for extending the discussion or focusing the ongoing conversation toward a convergence. Second, paraphrase also changes the non-scientific terms or phrases used by the students and "re-voices" (O'Connor & Michaels, 1993) them into a more appropriate expression. This is extremely useful for supporting students having difficulty with scientific language, particularly bilingual students or emerging language learners.

In the last excerpt (Excerpt 3.8) discussed under probe, the example also showed how the teacher paraphrased Stella's explanation for why a solid is unable to conduct electricity. In line 5, the teacher summarized the "two things" that were previously mentioned by Stella: (a) no free-moving ions and (b) strong electrostatic forces of attraction. In summarizing Stella's inputs, notice that the teacher had also re-voiced her utterance from "they are bonded together" into "strong electrostatic forces of attraction" (line 5), which is a denser and more scientifically accurate expression. Here are some more examples of re-voicing:

EXCERPT 3.9

#	Speaker	Utterance	Move
1	Teacher	Can someone explain why the paper clip does not fall?	I
2	Jason	Because it sticks to the magnet	R
3	Teacher	That's right, because it is **attracted to** the magnet	F

EXCERPT 3.10

#	Speaker	Utterance	Move
1	Teacher	So how do some animals conserve their energy during winter?	I
2	Celine	They go into deep sleep	R
3	Teacher	Okay. They **hibernate** or they go into **hibernation**	F

Reflective Toss. A "reflective toss" is a discursive technique used by a teacher to redirect the responsibility of evaluating or commenting on a prior response back to the same student who provided the response or to a different student (Van Zee & Minstrell, 1997). Through the repeated occurrences of IRE or IRF exchanges, many students have become accustomed to the teacher having to evaluate or comment on everything that was said in a class discussion. From this expectation, a teacher can sometimes provide too much guidance and consequently reduce the opportunity for students to take ownership of the thinking process in the discussion. Therefore, a reflective toss is a useful tool way to disrupt the social norm in classroom discourse and redirect the thinking and evaluation back to the students. Reflective toss can be easily achieved through questions like *what do you think of her answer?* Or *do you agree with what she just said?* It can also be used when a student has a question to ask. Before answering the question, the teacher can use reflective toss to direct the question to other students first.

The excerpt below shows a good example of a teacher pressing Melanie to reflect on her answer through reflective toss. Melanie has just presented her answer to the class after another student (Josephine) has presented hers. Instead of comparing and evaluating both their answers, the teacher asked Melanie to compare it herself in line 1. Through further prompting in lines 3 and 5, this resulted in Melanie noticing that her answer was missing a detailed explanation for the molten state. Then in line 9, the teacher did another reflective toss, but this time she redirected the evaluation to the rest of the class in order to engage them in the conversation:

EXCERPT 3.11

#	Speaker	Utterance	Move
1	Teacher	Melanie ... **when I ask you to compare** your answer to her (Josephine's) answer, is there a difference?	I
2	Melanie	About the same	R
3	Teacher	About the same. So that means there is still a little bit not the same. **So what is not the same?**	F
4	Melanie	Her answer is more detailed than mine.	R
5	Teacher	Her answer is more detailed than yours. That means something is missing from your answer. **What is missing from your answer?**	F
6	Melanie	The molten state thing	R
7	Teacher	Okay, what about the molten state thing? You didn't mention at all. Did you mention? Only a little bit. Can you read out what you have for the molten state?	F
8	Melanie	Potassium chloride in molten state, however, have free-moving ions as they are not held together by electrostatic forces of attraction.	R
9	Teacher	Alright. (*To the class*) **So what is it that she has missed out girls?** She says that in the molten state the ions are free to move about to conduct electricity.	F

Constructive Challenge. When a student gives a wrong or inappropriate answer, it is sometimes necessary to provide immediate corrective feedback. But instead of giving a direct evaluation, a teacher can pose a question or point to a contradiction in order to challenge the students' reasoning and make them reassess their answers. Chin (2007) calls this technique a "constructive challenge." When used appropriately, constructive challenge can not only reduce a student's embarrassment for giving an incorrect answer, but it also allows the student to realize why the proposed answer is not accepted instead of being told directly.

If you recall in Excerpt 3.2 with the "drinking with two straws" discussion, Cindy's explanation included an incorrect response that the pressure in the mouth (P_M) is greater than the pressure in the straw (P_S). Instead of correcting this response, the teacher first used reflective toss to get the rest of the class to comment on it. However, when there was no response after six seconds of wait time, the teacher changed his question to use a constructive challenge, as shown in the following excerpt:

EXCERPT 3.12

#	Speaker	Utterance	Move
1	Teacher	She said P.M. is greater than P.S., the pressure in the mouth is greater than the pressure in the straw, **what do you think?** (*6 seconds wait time*)	F
2	Teacher	Or **what do you guys think will happen if the pressure in the mouth is greater than the pressure in the straw?**	F
3	Rod	(*gestures a blowing action*)	R
4	Teacher	So Rod says this (*mimics Rod's action*). What does this mean, Rod?	F
5	Rod	Blow bubbles into the water	R
6	Teacher	So, actually if there's higher pressure in the mouth, then the air will actually move from the mouth into the straw inside the cup. It would become like this (*demonstrates blowing bubbles into water*). **But did we see that?**	F

The constructive challenge in this case started with the assumption that supposed Cindy was correct that the pressure in the mouth was greater than the pressure in the straw. The teacher then asked the class in line 2 what would happen if this statement was true. By following the logic, the students would have observed bubbles blowing into and out of the water. As this was not observed, therefore they could conclude for themselves that the statement was not true.

Fostering Student Interaction with Collaborative Questioning

While teacher questioning can shift the class interaction toward more dialogic, another approach is to let students develop good discourse strategies to mediate their interaction among themselves. Obviously, one of the most important things in this area is to allow more time and opportunities in the classroom for students to engage in small group discussions. There are also many cooperative strategies and techniques that teachers can use to support group learning, including well-known strategies like think-pair-share, jigsaw, or round-robin (Hmelo-Silver, Chinn, O'Donnell, & Chan, 2013; Strebe, 2017). These strategies are useful in providing broad organizational structure, turn-taking, and ground rules for collaboration. However, they do not address the specific discursive moves that many students lack in talking to one another (Chin & Osborne, 2010). This is where a number of discourse strategies highlighted in this section can be useful to supplement group discussion and collaborative learning.

It is common knowledge that the kind of talk during group discussion tends to be uncooperative, off-task, and unproductive (Mercer, Dawes, Wegerif, & Sams, 2004). Part of the reason is because students do not have the experiences and expectations of what a good discussion looks like. They are also given little explicit guidance on how to talk effectively in groups. This is where teachers should provide explicit instruction and modeling to students on how to raise collaborative questions and communicate ideas to one another in a group. As Chin and Osborne (2010) argue, students need to learn how to "ask questions that would help them become aware of what they do not understand, compare the strengths and weaknesses of competing ideas, recognise any inconsistency or faulty reasoning, formulate and test hypotheses, evaluate the evidence that supports or refutes the hypotheses, and generate alternative explanations or ideas that are more viable" (p. 235). In this regard, the various follow-up questions and moves discussed earlier are not exclusive to teachers' use only as they can also be used to engage students in a dialogic conversation. In particular, probe, extend, and constructive challenge are useful follow-up moves for students to learn from the teachers.

One useful recommendation for students to learn and practice their questioning skills is to pair their questioning with a round-robin cooperative strategy. First, following a round-robin, allow each member in the group a turn to present his or her ideas on the discussion topic or question. Repeat this for the next member until everyone has presented their ideas. However, in addition to this turn-taking procedure from a round-robin strategy, after a student has voiced his or her ideas, get another student to ask a follow-up question and keep asking until no more questions can be asked or until every member is satisfied with the collective idea that has been discussed. In effect, the student asking the questions acts as a teacher in facilitating an I-R-F-R-F interaction with the presenting student. To support students in asking appropriate follow-up questions, teachers can prepare prompts or a checklist to help students select and frame their questions to their peers. An example of a list of prompts is shown in Table 3.2.

If we recall the earlier conversation in Excerpt 3.4 between two students – Shanti and May, we can see that May was unconsciously using probe (line 2) and constructive challenge (line 4) as she questioned Shanti's idea that, "this P.S. (P_S) is bigger." Such student questioning is something to be encouraged. In

TABLE 3.2 Prompts for student questioning for various follow-up moves

Follow-up Moves	Some Prompts for Student Questioning
Probe	Why?
	How do you know that?
	What is your evidence?
	What is the concept or principle behind your reasoning?
Extend	So what?
	Can you elaborate further?
	How does that answer the question?
	How does this explain . . . [an observation or phenomenon]?
Constructive Challenge	But then you have . . . [a different observation or factor]?
	What about . . . [a missing link or factor]?
	Have you considered . . . [other relevant factors] . . . ?
Paraphrase	Are you saying that . . . ?
	Can you rephrase that in another way?
Reflective Toss	What do you all think of his/her answer?
	Does everyone in the group agree?

Shanti and May's conversation, unfortunately, this peer-to-peer question and answer lasted only four turns before they decided to ask their teacher for help. Using the prompts in Table 3.2, we can perhaps guide May to keep asking follow-up questions until they reach a better understanding of their developing explanation instead of giving up so quickly. In addition, when May has finished her questioning, both students should swap their roles following a round-robin method. Thus, May will have to present her thoughts on the issue while Shanti takes on the role to ask follow-up questions using the prompts in Table 3.2.

In providing explicit instructions for students to ask good follow-up questions, a clarification of several terminologies, or metalanguage, may be necessary. For example, words like evidence, principle, reasoning, observation, phenomenon, and theory found in Table 3.2 should be clarified with students. These are meta-scientific terms with precise meanings that most students are not familiar with. By introducing this metalanguage, as we will explore further in Chapter 6, they provide a useful language for students to frame their questions. An interesting example is shown below in Excerpt 3.13 which occurred during a teacher-led discussion. Although the discussion was led by a teacher, the usual IRE/IRF triadic structure was disrupted when a student, Nguyen, gave a fairly comprehensive explanation to the question, "Why does hot air rise?" Nguyen's long explanation was then followed by three students who voiced their objections to the explanation. These objections from the students are good examples of constructive challenges, as we see here:

EXCERPT 3.13

#	Speaker	Utterance	Move
1	Nguyen	I was thinking of the comparison between the cool air and the density of hot air. So when you compare, hot air will rise because lesser density. Normally when you compare, you will say the denser one will sink … I was thinking of comparing the normal air density and the hot air density. So when you compare, normally you always say the denser one will sink. So since hot air rises, means the density of the hot air is lower than the density of the normal air. Since normal air is having, is denser. So it will sink. So when it sinks, the hot air rises due to the density is lower. It's like you compare with a liquid and a cork …	State
2	Rhoda	**What's the principle?**	Question
3	Teacher	Yah, it's already stated here	Answer
4	Jihoon	**But it's not related to …**	State
5	Farid	**It's not linked to any principle**	State

The example in Excerpt 3.13 illustrates an outcome that had developed from two pedagogical interventions. First, the students' constructive challenges were made possible because they had previously learned the metalanguage "principle" in relation to the genre of a scientific explanation (see Chapter 6 for more on this metalanguage). Second, the teacher had frequently let the students engage in group discussions and practice asking questions among themselves. Thus, he had created a conducive culture where students were not afraid to ask each other questions or voice their constructive challenges. This conversation provides an evidence of student learning through a change in their discourse as they appropriated the dialogic questioning and metalanguage first introduced by their teacher. In addition, this example also illustrates that student questioning is not restricted to peer discussion. Once students have developed the skills and habits of asking questions in small groups, they should be encouraged to raise their questions during whole-class teacher-led discussions. This is a good way to disrupt the authoritative IRE or IRF interaction pattern and make the classroom interaction more dialogic.

Fostering Student Interaction with Argumentation and Group Representation

Besides student questioning, another way to foster dialogic interaction among the students is to use argumentation and group representation. The ideas behind these two discourse strategies will be further elaborated in Chapters 6 and 7 respectively. For now, I will briefly outline how they connect to the interactive patterns in student-student dialogue.

The essence of a dialogic interaction is there are at least two points of view or voices interacting in a continual chain of speech communication (Bakhtin, 1981). At times, dialogic interaction is difficult to achieve in a science classroom because there is often a "correct answer" from the scientific point of view. Thus, in a group discussion, most students are usually seeking for an answer through the discussion rather than engaging in a dialogue to present or defend a particular idea or point of view. This explains why many small group talks (even when they are cooperative) tend to revolve around sharing brief and factual information as well as agreeing or acknowledging a previous response. By comparison, students challenging one another's ideas and claims are much rarer. One way to overcome this challenge is to design instructional activities for students to engage in argumentation, which involves the use of evidence to persuade one's peers in favor of a particular claim or position (Erduran, Simon, & Osborne, 2004). For argumentation to occur, there must be at least two competing claims or positions to argue about (Osborne & Patterson, 2011).

To promote an argumentative talk, teachers will therefore need to develop questions that allow sufficient discussion from two or more distinct or conflicting points of views. The open-ended questions discussed in an earlier section are useful in this aspect. Some of these questions are "Does light travel forever?", "What kills the dinosaurs?", and "Is potato a root or stem?" Based on the question, students should be allocated or assigned to defend one of the positions. Using a debate as an activity for small group or whole-class discussion, the ensuing talk will be much richer and more dialogic because each student will have a personal stake to defend a particular position, as we will see in Chapter 6.

Separately, student-generated group representation such as group drawing, poster, or physical artifact is also another useful way to promote collaborative and dialogic talk. This approach utilizes another semiotic mode besides verbal language for students to build and assess their understanding. In particular, semiotic modes that draw on concrete and pictorial representations (e.g., diagrams, tactile objects) tend to provide more creative space for students to generate their own ideas as compared to symbolic representations (Tang, 2016b). Thus, the use of group representation is a useful strategy as it combines the benefits of using a non-verbal mode of representation and collaborative talk. In particular, as students are given opportunities to discuss their ideas verbally, they also record and make explicit their ideas on a permanent medium (e.g., paper, screen), which then serves as a public space to prompt further dialogue (Park, Chang, & Tang, in preparation). We will examine this in more detail in Chapter 7.

Summary

The interaction pattern of classroom discourse is the first of five discourse patterns that are introduced in this book. As it provides a social structure for all conversations to occur, it is arguably the most fundamental discourse pattern to be observed in the classroom. In education research, it also has the longest history since the days when the IRE pattern was documented by classroom ethnographers in the 1970s. Since then, there has been a wealth of literature written in this area revolving around topics like classroom interaction, discursive roles and identities, speech act, conversation analysis, questioning techniques, and collaborative discourse. There is a lot more I can say concerning these topics but this is not possible within the space of one chapter. Readers who are interested in these topics are encouraged to explore further based on the basic ideas and references I have provided in this chapter.

In this chapter, I have briefly described four major types of interaction pattern in a classroom – IRE, IRF, teacher talk, and student dialogue, and how they operate with complementary roles to fulfill various instructional purposes in a typical science lesson. With respect to the corresponding discourse strategies, I have focused mostly on strategies that deliberately promote dialogic interaction through IRF and student dialogue. This is because science classrooms tend to be authoritative and overly teacher-dominated. Thus, there is a real need for teachers to be aware of the implicit discourse strategies they use to maintain teacher control and to explore other strategies that can encourage student involvement and thinking. These discourse strategies which were made explicit in this chapter include dialogic questioning, follow-up moves, and collaborative questioning.

As interaction patterns are fundamental to all social conversations, the basic ideas and strategies presented in this chapter are not exclusive to science and are also applicable to all academic subjects, including language, mathematics, and humanities. This versatility of interaction pattern and its corresponding discourse strategies is, however, also its major limitation when it comes to understanding the "science" in science classroom discourse. Interaction pattern does not tell us anything about the stuff that people talk about through their interaction. This is where we need to examine other discourse patterns in subsequent chapters that will inform us about the content (Chapter 4), narration (Chapter 5), practices (Chapter 6), and representations (Chapters 7 and 8) of science. We will start by unpacking the content of science through the lens of thematic discourse pattern in the next chapter.

Note

1 This method of analyzing adjacency pairs according to a system of giving/demanding information or goods and services is informed by the role of speech functions in Halliday's SFL. Speech function is an interpersonal metafunction of language to help speakers interact and negotiate with one another in a dialogue. This method could also be used to analyze the teacher's whole-class discussion. Thus, an IRE is usually a Question-Answer-Statement exchange unit while an IRF is usually a Question-Answer-Question exchange unit.

4

USING DISCOURSE TO

Build and Assess Scientific Content Knowledge

In Chapter 3, we establish that classroom talk is a social conversation with various interactional structures (e.g., IRE, IRF, monologic lecture). These structures are discursively produced through the words we say, along with other non-verbal cues, in the form of questions, answers, statements, and acknowledgments, to name a few. However, as we have theorized in Chapter 2, words do not just form interactional structures through their interpersonal metafunction, they also have an ideational metafunction to construct some kind of "content" between two or more people (Halliday, 1994). For this reason, the interaction and content in any classroom discourse always occur simultaneously. Thus, they form a basic dyad in classroom discourse that can be examined through the lenses of interaction pattern (Chapter 3) and thematic pattern (this chapter).

As a brief illustration, the following transcript was examined through the lens of interaction pattern in the previous chapter. In particular, this short dialogue between the teacher and a student, Zac, was characterized by two consecutive IRE exchanges. In the first exchange, Zac gave the incorrect response to the teacher's question in line 1 and thus it prompted the second exchange where Zac was given a second chance to try again in line 3:

EXCERPT 4.1

#	Speaker	Utterance	Move
1	Teacher	Zac, what is the S.I. unit for force?	I
2	Zac	Newton per square meter	R
3	Teacher	Force. Not pressure	E/I
4	Zac	Oh. Newton	R
5	Teacher	Newton, alright. So the unit for force is Newton	E

We have discussed how an interaction pattern provides a structure and technique for science teachers to ask questions and provide feedback, but we did not address the question of how do teachers know what question or feedback to ask or give in the interaction. In the above exchange, besides the IRE interaction, there was also a content being constructed simultaneously. A simplified thematic pattern of this content is visually represented in Figure 4.1. This pattern consists of two semantic relationships: attributive and identifying. Semantic relationships and thematic patterns will be further elaborated in this chapter. For now, I want to highlight that this thematic pattern was not only built up by the teacher and Zac over two IRE exchanges, it is also repeated over and over again in many places beyond this classroom. In fact, it is precisely the reoccurrence of this thematic pattern that determines the correctness of Zac's answer in line 2 and consequently prompted the teacher to correct his response and start the next IRE exchange. In other words, the content of science (via thematic pattern) and the interaction among the participants (via interaction pattern) are always mutually dependent. One cannot occur without the other.

Science teachers tend to understand and talk about the content of science through notions like vocabulary and concept. Most science curricula around the world also prescribe a standardized list of vocabulary words and concepts that students need to learn by the end of every academic year. However, talking about vocabulary and concept is not very useful in terms of helping students learn science content. Every science teacher knows that learning science is more than just mastering a list of keywords. Emphasizing vocabulary tends to produce students who can regurgitate important scientific words and definitions without a good understanding of what they mean. The notion of concept, on the other hand, tends to invoke among science teachers something deeper than just vocabulary. But other than serving as a label for denoting some recurring ideas that students need to understand, few educators can articulate clearly what is a scientific concept, and more importantly, how do we teach a concept to students?

Following the core ideas of discourse pattern and discourse strategies in this book, this chapter elaborates the thematic pattern that is associated with the content of science. Thematic pattern examines the connections among the words we say or write in our communication, and pieces together the semantic

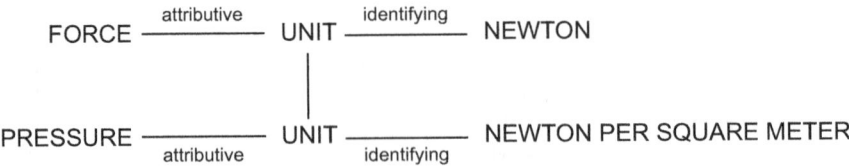

FIGURE 4.1 A thematic pattern of the content communicated in the exchange

relationships that construct the ideational meaning (or content) in the communication. Unlike the notions of vocabulary and concept, thematic pattern provides a more explicit and transparent way to analyze science content matter based on the actual words[1] we hear, say, read, and write in the classroom. In the first half of this chapter, I begin by discussing how we learn the meaning of a word, before going into a more technical discussion on semantic relationship, thematic pattern, and how these two ideas are related to scientific concepts. With an understanding of thematic pattern, I introduce several discourse strategies related to thematic pattern in the second half of the chapter.

Thematic Patterns of Science Content

Meaning of a Word: Through Semantic Relationship

If someone asks you what does a particular word mean, how do you usually respond? For some words, we can point to an object (e.g., burette, pendulum), show a picture, or mimic an action, and say this is what the word means. This approach is, however, limited to firstly, only the names of objects and processes that we can see, and secondly, there have to be material things nearby that one can point at or show. For most words, we have to say what the word means by *saying more words*. This can range from a few words such as a synonym to a string of words that provide a definition or describe the word's attributes. Whether it comes from a person or a dictionary, other words are used to provide the meaning of a word. It is important to distinguish that the meaning is not contained in the words themselves, but rather, the person hearing or reading those words have to construct the meaning through the *semantic relationships* among those words. As discussed in Chapter 2, this is what we mean by saying that meaning is made or constructed through the choices of words in a language (Halliday, 1978), and it applies whether we are communicating to someone or thinking to ourselves (as inner speech; Vygotsky, 1986).

To further illustrate what semantic relationships are and their role in making meaning, consider the following example of a written text about the animal kingdom. For reasons that will be made known later, I have masked the names of the animals and replaced them with some arbitrary names:

> The animal **Jeeta** belongs to a biological family called **Heebra**. Other examples of **Heebra** are **Flo**, **Himmo**, and **Cham**. Although quite similar in shape and size to the **Jeeta**, the **Bew** is actually from the **Pauline** family. The **Lima** and **Pane** are two different types of **Jeeta**.

Suppose you are a student who is reading this text, which words are the new vocabulary to you? How do you make sense of these words in this text?

It will be useful if you take a few minutes to interpret this text by drawing a tree diagram to show the connection among the words highlighted in bold. When you have finished, compare your diagram with Figure 4.2.

If you have created something similar to Figure 4.2 by reading the text passage, how do you make the connections among those words even though you do not have any prior knowledge of those words? The cues that allow you to form the connections come from words like "belongs to," "examples of," "from the … family," and "types of." These are different ways of saying the same thing. Thus, saying that "**Jeeta** belongs to **Heebra** family" and "Examples of **Heebra** are **Flo**" are similar because *semantically*, they construct a relationship that subsume **Jeeta** and **Flo** under the larger category of **Heebra**. These are examples of one type of semantic relationships called *classifying*, which classifies entities into a hierarchical relationship. This is an example of how people make meaning from the words they read, or to be more specific, through the semantic relationships among words.

The above example is not hypothetical even though the words in bold may sound foreign to you. It is actually based on a real text that describes the different species of fish and mammal, as follows:

> The animal **shark** belongs to a biological family called **fish**. Other examples of **fish** are **swordfish**, **goldfish**, and **tuna**. Although quite similar in shape and size to the **shark**, the **killer whale** is actually from the **mammal** family. The **great white shark** and **whale shark** are two different types of **shark**.

Notice that the only thing I have altered in the first text was to replace the familiar names (e.g., shark, swordfish) with arbitrary labels (e.g., Jeeta, Flo),

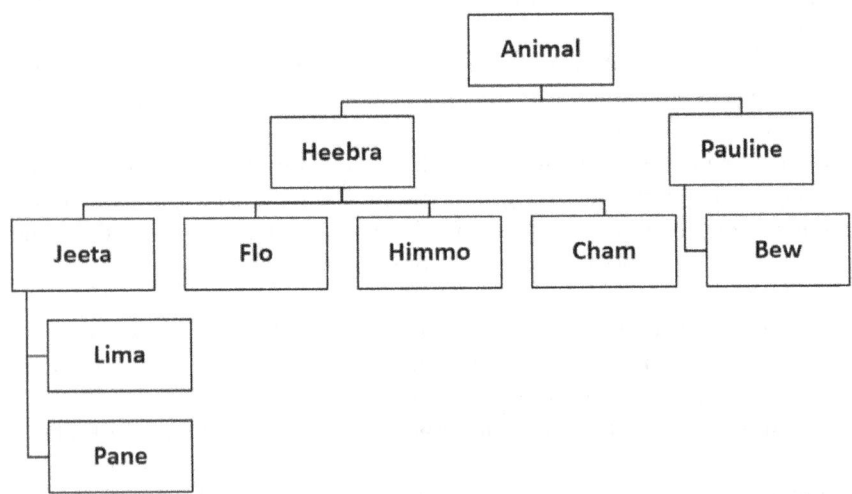

FIGURE 4.2 Relationship among the new words in the text passage

while maintaining their semantic connections. The reason is twofold. First, it is to illustrate that most English words in both their phonetic and written form have little resemblance to the actual things they represent. Thus, I can use any arbitrary word to replace the more familiar words. For instance, the words "Jeeta" and "shark" do not look or sound anything like the sea creature. This illustrates that a word is just a symbol and its meaning does not come from the word itself, but is derived based on conventions from its semantic connections to other words. According to SFL theory, as explained in Chapter 2, the meaning at the discourse semantics level is realized from a pattern of lexicogrammar (words), which is itself also a pattern of phonology (sound) and graphology (symbol).

Second, this example also illustrates how we learn new vocabulary. Most adults will have some prior knowledge that shark is a subclass of fish. But at some point in their past, they must have learned this knowledge somewhere for the first time; perhaps by reading a book on fish, participating in a family conversation in the aquarium, or watching a documentary about sharks. The circumstance may not be a formal statement such as, "shark is a type of fish," but rather through everyday conversations. For example, a child looks at a shark in an aquarium and remarks, "Look at that giant fish!" This is then followed by a reply from a parent, "That's a shark!" The circumstance can also be looking at a picture of a shark in a poster with the title, "The fishes of Western Australia." Regardless of the circumstances, the meaning of shark is always made through the semantic relationships among words, symbols, and objects in the physical world. (A similar set of semantic relationships also exists for non-verbal language, such as diagrams and gestures, and this will be further explored in Chapter 8.)

While this example shows how we make meaning of new words like Jeeta and Flo (or shark and swordfish), it only illustrates one type of semantic relationship. We will need more words (and other symbols) to construct other types of semantic relationships in order to give a more complete description of the natural world we live in. This is where we need to know a range of semantic relationships.

Semantic Relationships

Halliday (1985) had identified about a few dozen basic semantic relationships that mostly cover the range of meanings we make through the ideational metafunction of verbal language. There have been different ways to classify and organize these semantic relationships into a coherent framework to suit various purposes (e.g., Lemke, 1990; Martin & Rose, 2007; Unsworth, 2001b). Drawing from these sources, I have organized several semantic relationships typically found in scientific discourse into four different categories, as shown in Table 4.1. In this book, I simplify the range of semantic relationships to make it easier for science educators without a working knowledge of linguistics to identify for the purpose of understanding the thematic pattern of science classroom discourse.

TABLE 4.1 Classification of semantic relationships in verbal language

Category	Semantic Relationship	Example
Transitivity	Relational processes:	
	• Classifying	Shark *is a type of* fish
	• Composing	Matter *is made of* atoms,
	• Attributive	Whales *are* warm-blooded
	• Possessive	A moving car *has* kinetic energy
	• Identifying	This pressure *is called* the turgor pressure
	• Quantifying	The temperature *is* 25 °C
	Physical processes:	
	• Material	Salt *dissolves* in water, magnet *attracts*
	• Existential	There *is* no air in space
	Perceptual processes:	
	• Behavioral	We *observe* that there is more water vapor
	• Mental	Scientists *believe* in an asteroid collision
	• Verbal	Just now, you *say* that . . .
Circumstantial	Location	There is no air *in space*
	Time	The turgor pressure is high *at the start*
	Manner	The car stops *slowly*
Taxonomic	Hyponym	*Mammal – whale, human, dog*
	Meronym	*Atom – proton, neutron, electron*
	Synonym	*Accelerate – speed up*
	Antonym	*Increase – decrease*
	Scale	*Tightly packed – close together – scatter*
Logical	Addition	*And, in addition, furthermore*
	Comparative	*However, while, but*
	Causal	*Therefore, so, because, consequently*
	Temporal	*When, after, then, subsequently*
	Conditional	*If, then, unless*

The first category of semantic relationships revolves around what Halliday (1985) calls *transitivity*, which is a grammatical system that relates our ideas and experiences into units of meaning. In the English language, the clause functions as the basic unit of ideational meaning that constructs a particular event or sense of experience (Halliday, 1985). A clause must minimally consist of a verb and at least a noun. The nouns or noun phrases (e.g., light, air) represent the things or entities in the experiential world called *participants*. There can be many participants within a clause, but there must be a main participant (called a *medium*) without which the clause cannot exist. The verb in a clause represents the *process* that relates the participants together. According to

Halliday (1985), there are three major process types that roughly correspond to the experiential domain of (a) actions in the physical world of doing, (b) abstract relations among entities of being, and (c) perceptions in our consciousness (of sensing). These three process types therefore constitute three sets of semantic relationships within the category of transitivity relations, namely physical, relational, and perceptual processes respectively.

Physical processes, consisting of material and existential, highlight the material action or basis of a participant. Not surprisingly, as science is the study of the physical world, scientific language is rife with physical processes. Relational processes set up some kind of relationships between two separate entities. We have seen an example of a *classifying* relationship in the earlier fish example. Other types of relational processes include *compositional, attributive, possessive, identifying,* and *quantifying.*[2] Lastly, perceptual processes deal with human consciousness and actions, which include *behavioral, mental,* and *verbal* processes. As the propositional content of science typically omits human involvement and perception, perceptual processes are not very common in scientific texts as opposed to literary, humanities, and social science texts. However, as we will see in Chapter 5, perceptual processes are commonly used in our metadiscourse – a discourse of our discourse. Thus, perceptual processes function more as a commentary to our own talk rather than building the propositional content of science.

The category of *circumstantial* relation provides further information about things and events in the world in terms of the location, time, and manner. Circumstantial relation is typically realized through prepositional phrases (e.g., *at, by, during, for, of, on, in, under*) and some adverbs (e.g., slowly, away, toward). As such, it is often found in the peripheral of a clause, in comparison to transitivity relation that occupies the central nucleus of a clause (Martin & Rose, 2007). What this means is that while a clause can exist without a circumstance, it cannot exist without a transitivity relation (realized through verb and noun phrases). For example, in the clause "there is no air in space," it is possible to omit the prepositional phrase "in space" while still maintaining the linguistic structure of the clause, but not the other way round. Although circumstantial relation is not mandatory to make a clause from a linguistic point of view, this does not mean that its semantic meaning in terms of location, time, and manner is not important from a content point of view.

Taxonomic relations are quite similar to relational processes in terms of establishing some kind of abstract relationship between two entities. However, I make a distinction between these two categories because relational process (as a transitivity) is visible within a clause while taxonomic relation is either embedded within a phrase or scattered across clauses. For example, the phrase "kinetic energy" implies it is a *hyponym* of energy, similar to a classifying relationship. The phrase "electrons in the atom" implies electron is a *meronym* of atom, which is similar to a composing relationship.

Another way that taxonomic relation can be formed is when related words are used throughout the text. For example, words like "shark, tuna, salmon" and "fin, gill, tail" are related as different types of fish (co-hyponym) and parts of a fish (co-meronym) respectively. Other types of taxonomic relations are *synonym* and *antonym*, which are words with similar and contrasting meaning respectively. Another taxonomic relation is *scale*, which is similar to *antonym*, but instead of a binary opposition (e.g., good, bad), there is a range of meanings (e.g., excellent – good – fair – bad – appalling). All these taxonomic relations require the reader to have some prior knowledge of the vocabulary used, as opposed to transitivity relations which can always be read from the grammar in the text.

The last category is logical relations, which establish connections across successive clauses and sentences through conjunctions such as a*nd, in addition, furthermore* (additive); *however, while* (comparative); *therefore, so, thus, because* (consequential); *when, after, then, subsequently* (temporal); *if, then, unless* (conditional). Logical relations are crucial to building the sequences in many genres. For example, information reports are filled with additive and comparative relations while explanations have more temporal and consequential relations.

Thematic Pattern – A Pattern of Semantic Relationships

Semantic relationship is fundamental to the way we make meaning with a symbolic language. We intuitively form semantic relationships in our mind when we hear or read the words in a familiar language. When the meaning of a word involves one or two semantic relationships with several words, it is not difficult to understand the idea behind that word from its semantic relationship with other words. For example, the word "scarce" can be easily described to a child by saying it is similar to (synonym) "little" or opposite (antonym) to "many." However, when an idea involves a web of many semantic relationships, this can become very challenging for new learners to understand. Unfortunately, most scientific concepts are like that.

Scientific concepts are commonly accepted ideas about the physical world at a higher level of abstraction. What makes a concept an abstraction is not due to any inherent nature of the idea, but really due to a complex and hierarchical network of words associated with the concept in the sense that its meaning depends on our prior understanding of other concepts, which are themselves also a network of words. To understand what scientific concepts are and how they work, it is useful to introduce the idea of thematic pattern, which Lemke (1990, p. 12) defines as "a pattern of semantic relationships that describes the thematic content."

Consider this conversation between a chemistry teacher and her class of ninth-grade students. In the transcript, I have indicated the semantic relationship for every one or few lines that construct a semantic unit of meaning.

EXCERPT 4.2

#	Speaker	Utterance	Semantic Relationship
1	Teacher	Now next. Why do I choose to use fractional distillation rather than simple distillation?	fractional distillation − [material] − use − [conseq] distillation − [hyponym] − simple & fractional
2	Teacher	Can I use simple distillation?	simple distillation − [material] − use
3	Jenny	No	
4	Teacher	From what you recall, simple distillation we use it to separate what, ah?	distillation − [material] − separate
5	Melanie	Water	water − [attributive] − boiling
6	Su Fen	Boiling point more than 25	point − [quantifying] − > 25
7	Teacher	Boiling point more than 25	
8	Teacher	Apart from that? What kind of mixture do I use? Or maybe you give me one example, we'll look at that example	mixture − [classifying] − seawater − [composing] − salt & water
9	Rhoda	Salt and water	
10	Teacher	Salt?	
11	Mandy	Seawater	
12	Teacher	Ah, okay. You separate your seawater into salt and distilled water. Correct or not?	separate − [material] − seawater − [composing] − salt & water
13	Teacher	Okay, so that is one way that you use simple distillation	simple distillation − [material] − use
14	Teacher	But why is it that we use fractional distillation here? What's the main reason?	fractional distillation − [material] − use − [conseq]
15	Thila	Boiling point	Liquid oxygen and nitrogen −
16	Teacher	Yah. Take a look at the difference in their (liquid oxygen and nitrogen) boiling point. What do you notice about the difference in boiling point?	[attributive] − boiling point − [attributive] − very close
17	Thila	They're very close	
18	Teacher	Ya, they're very close	
19	Teacher	Alright, so first of all, liquid air is a mixture of liquids, alright, having similar or, or, or, close boiling points	mixture − [classifying] − liquid air [composing] − liquids
20	Teacher	Alright, so therefore I use fractional distillation	fractional distillation − [material] − use − [conseq]
21	Teacher	Okay, and for simple distillation, we use it for soluble solid-liquid mixtures	solid − [co-hyponym] − liquid

In the beginning of this conversation, the teacher introduced an important semantic relationship that revolved around the word "use." This is a material process that involves two main entities – fractional and simple distillation. This semantic relationship was repeated many times throughout the conversation (e.g., lines 2, 4, 13, 14, 20, 21), which would indicate its importance and overarching purpose for the conversation. Furthermore, this semantic relationship was made together with the main question of "why" in lines 1 and 14 and the conjunction "therefore" in line 20. We will come back to how this juxtaposition sets up a logical (consequential) relation that provided the main rhetoric in this explanation to address the "why" question. For now, we need to examine the rest of the thematic content that was constructed by other semantic relationships.

Another material process besides "use" is the physical action signified by the word "separate" in lines 4 and 12. There are two entities joined by this process. As seen in line 4, the two entities are "simple distillation" and an unknown "what" in the teacher's question. The teacher's follow-up question in line 8 provided further clues that this unknown entity is a "kind of mixture." Finally, in line 12, the unknown entity was established to be "seawater," first proposed by Mandy in line 11. A similar relationship involving the material process of "separate" was also observed for "fractional distillation" from line 14 onwards. Although there was no explicit mention that [fractional distillation – separates – liquid air], this semantic relationship can be implied if we examine the thematic pattern and compare the process between simple and fractional distillation, which we will do so later.

For simple distillation, we saw that the material process "separate" was split into two parts, first involving "simple distillation" in line 4 and later with "seawater" in 12. But what semantic relationships were formed in between from line 5 to 11? In these exchanges, the teacher and several students discussed the nature of seawater, and this was achieved through several relational processes. The first is a classifying process that established "seawater" is a type of "mixture" from the question-answer pair between lines 8 and 11. This classifying process was also repeated for fractional distillation when the teacher said, "liquid air is a mixture of liquids" In line 19. The second relational process is compositional where "salt" and "water" are parts of seawater, as shown in lines 9, 11, and 12. This process was not repeated for fractional distillation, and thus it has to be implied by comparing with the overall thematic pattern.

The third relational process is attributive which constructed boiling point as one of the properties of salt and water for simple distillation, and liquid oxygen and nitrogen for fractional distillation. This was explicitly mentioned for fractional distillation in line 16 through the phrase "the difference in their (liquid oxygen and nitrogen) boiling point." For simple distillation, this was indirectly implied through Melanie's and Su Fen's incomplete answers in lines 5 and 6. The last relational process is quantifying, which ascribes a numerical value to the difference in the boiling point in lines 6–7 ("more than 25") for simple distillation. A similar relationship is found for fractional distillation in lines 17–18, although in this case, a student said "very close," which forms an attributive relationship, instead of giving a numerical value. Nevertheless, this answer was accepted by the teacher in line 18 because for the purpose of this explanation, it makes no difference if she had said "very close" or "less than 25."

Besides these transitivity relations (both material and relational processes) that are visible in the conversation, there are two subtle hyponyms that are implied and not obvious. The first hyponym was found in line 1 where the phrases "fractional distillation" and "simple distillation" imply that both are types of distillation. In line 21, there was another hyponym where solid and liquid are different types of matter. This was not overtly stated and it was assumed that students in Grade 9 would have this prior knowledge. The term "liquid air" would probably cause a lot of confusion as many people mistake air as a type of gas (a hyponym), thus air cannot be a type of liquid. However, air is not a type of gas (a hyponym). It is a mixture of nitrogen and oxygen (a meronym) and these elements can exist in a liquid state at low temperature. Thus, "liquid air" is a type of liquid and not a type of gas. This relationship is not easy to see unless one expands the hidden taxonomic relationships (hyponym and meronym).

After examining each of the major semantic relationships in detail, we are now ready to piece together the thematic pattern of this conversation. For analytical and communication purposes, it is sometimes useful to draw a visual representation of a thematic pattern. In this visual representation, the main lexical items (or keywords) are capitalized and they are joined by lines to related lexical items. The labels on the lines indicate the semantic relationships. I use straight lines to indicate the semantic relationships that are explicitly mentioned in the conversation and dotted lines to indicate relevant relationships that are implied or assumed to be known. Horizontal lines indicate transitivity and circumstantial relations where the meaning is usually formed within a clause (e.g., distillation *separates* seawater). Vertical lines indicate taxonomic relations where the meaning is formed within a noun phrase (e.g., *fractional* distillation) or across clauses (e.g., water, liquid nitrogen, liquid oxygen). I also use arrows to highlight logical relations (e.g., *therefore*, I use fractional distillation). Based on this convention, the thematic pattern for this conversation is visualized in Figure 4.3.

Thematic Pattern and Scientific Concept

Thematic pattern can be represented in a visual diagram such as Figure 4.3 in order to show and analyze the "pattern of semantic relationships." From the diagram, one can see a corresponding pattern between simple and fractional distillation in terms of their similar and contrasting semantic relationships. Horizontally, the material and relational processes for both of them are almost

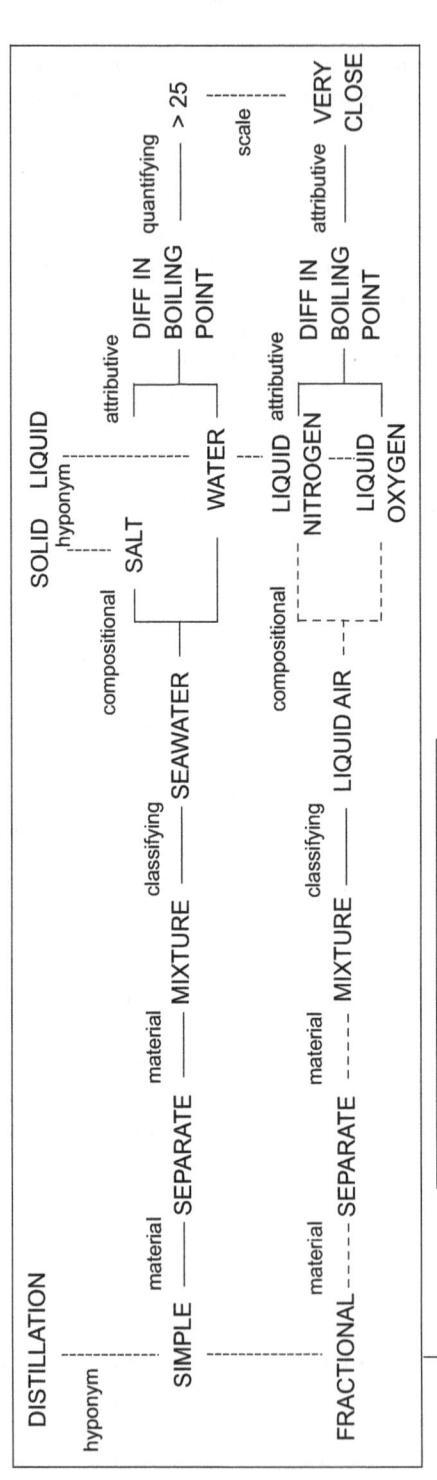

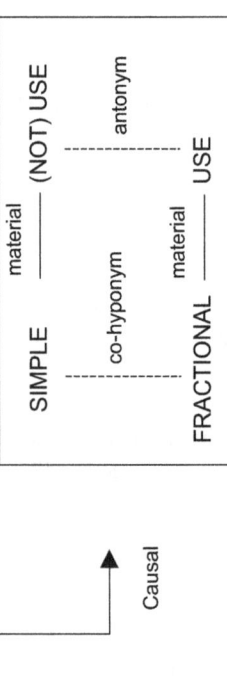

FIGURE 4.3 A thematic pattern of "simple and fractional distillation"

identical. Because of these similarities, this was how most people can piece together the missing semantic relationships even though they were not explicitly uttered. Earlier, I have taken note of two missing semantic relationships for fractional distillation: (a) [fractional distillation – separates – liquid air] and (b) [liquid air – comprises – liquid oxygen and nitrogen]. These missing links are represented as dotted horizontal lines in Figure 4.3. By noticing the similar semantic relationships for simple fractional distillation, most students could "fill in the blanks" to understand the same semantic relationships for fractional distillation as well. Thus, individual semantic relationships and thematic pattern often form a mutually reinforcing connection; just as many semantic relationships form a thematic pattern, the developing thematic pattern also becomes a conceptual structure to understand the underlying semantic relationships.

Just as the similarities are clearly shown in the thematic pattern diagram, the key differences between the use of simple and fractional distillation are also evident. The main point of difference is not between "seawater" and "liquid air" as lexical items, but between their corresponding relational processes. After all, seawater and liquid air are just specific examples of mixture and they can be substituted by other substances, which will still retain the overall thematic pattern. What is more crucial is the underlying semantic relationships. In this case, there are two semantic connections that explained the difference between simple and fractional distillation. The first connection involved the attribute or value of the "difference in boiling point," and this was a key focus that was developed through the teacher's IRE/IRF interaction with the students (lines 5–11 and 15–18). The second connection involved hyponym where simple distillation is typically used for "soluble solid-liquid mixtures" (line 21) while fractional distillation is used for liquid-liquid mixture. This connection was not well developed in the conversation, and it required the students to see the subtle hyponym where salt is a solid while water, liquid oxygen, and liquid nitrogen are all liquid. Although most ninth-grade students would no doubt have this prior knowledge, it is a different story whether they are aware of it and could activate this prior knowledge to make these connections *at the moment* when this discussion occurred.

Lastly, we can also see from the thematic pattern what constitutes the concept of distillation and how this was used as a rhetoric or "reason" (line 14) to explain the choice of using fractional distillation (lines 14 and 20) but not simple distillation (lines 2–3). In the beginning, the teacher first established an antonym (opposite) between "using fractional distillation" and "not using simple distillation." This antonym was then juxtaposed against the pattern of semantic relationships (consisting of several similar and contrasting connections as elaborated in the last two paragraphs) through a logical consequential relation that were repeated at several places, such as "why" (lines 1, 14), "reason" (line 14) and "therefore" (line 20). In other words, the concept of distillation and the internal logic of the explanation was formed and communicated through a network of semantic relationships consisting of material and relational processes as well as taxonomic and logical relations. This example illustrates that scientific concepts are not obscure ideas that exist as an abstraction in our mind, but they are formed through a web of many semantic relationships.

The thematic pattern drawn in Figure 4.3 is not complete as it is constructed based on a very brief conversation that lasted only 21 turns and a few minutes. It primarily focused on the semantic relationships that linked simple and fractional distillations to several lexical items such as mixture, separation, boiling point, solid, liquid, and liquid air. We could say that these lexical items are the prior concepts that students need to know in order to understand the concept of distillation. But what are these prior concepts other than also just another thematic pattern of semantic relationships? For any of these concepts, say mixture, the expression of its thematic pattern can be found in written curricular documents (e.g., textbooks) as well as other conversations that talk about mixture. Thus, if we expand every lexical item into its thematic pattern and join these patterns to the thematic pattern in Figure 4.3, you can imagine the larger thematic pattern which is going to be extremely vast and complicated. Although this pattern will be too large to be practically useful, it illustrates a theoretical point that a scientific concept is essentially a complex and hierarchical network of words arranged in a certain patterned way.

The last characteristic of scientific concept I like to elaborate is its shared and institutionalized nature. The thematic pattern shown in Figure 4.3 is not unique to this specific conversation in one classroom, but it is repeatedly occurring in most chemistry classrooms and textbooks on the same topic around the world. It is also not restricted to the use of the English language because the pattern would be the same if the participants had spoken Mandarin, Spanish, or French. Although every conversation is unique and there are differences in vocabulary and grammar, the basic underlying semantic relationships are more or less similar. This pattern is always repeated because there is a standardized way of talking in order to be recognized as making sense of the particular concept (e.g., distillation) according to the social practices of the scientific community. If any student deviates considerably from this pattern, someone else who is familiar with the pattern will recognize the inconsistency and react as such. In the beginning of this chapter, we saw an example of where the teacher corrected Zac's answer for the S.I. unit for force in order to maintain the thematic pattern as shown in Figure 4.1. As such, teachers, textbooks, assessment materials, and examination boards serve as gatekeepers to reinforce the institutionalized nature of scientific concepts, and they do so by maintaining the integrity and reoccurrence of the underlying thematic patterns made in science classrooms and examination settings around the world.

Thematic Discourse Strategies

The construction and interpretation of semantic relationships is an integral part of our human meaning-making experience. When teaching a concept, every teacher instinctively uses a range of discourse strategies to make various semantic relationships that relate to the thematic pattern associated with the concept, just as we saw how the chemistry teacher directed her students to talk about the concept of distillation. Just like interaction patterns, many of these thematic discourse strategies are implicitly carried out without much thought and deliberation. However, when we become aware and more conscious of these implicit discourse strategies, we can find ways to make them more explicit and consequently more effective in improving student learning. In this section, we explore several strategies that aim to explicitly support students in making semantic relationships as they learn scientific concepts. For the purpose of easy referencing, I name these strategies as follows: critical semantic connection, unpacking abstraction, and jumbled sequencing.

Mind the Gap: Critical Semantic Connection

As illustrated in the previous section, the teaching of scientific concepts requires an assemblage of semantic relationships to build up its thematic pattern. Quite often, some of these semantic connections were unintentionally omitted in our instruction, especially when we do not pay sufficient attention to the language used in our classroom discourse. In the last example, we saw two important semantic relationships were not mentioned in the talk on fractional distillation (represented as dotted lines in Figure 4.3): (a) fractional distillation separates liquid air and (b) liquid air comprises liquid oxygen and nitrogen. Most students might be able to piece together these missing links, provided that the overall thematic pattern was systematically presented (as was the case in the distillation example). However, if the thematic pattern was not well developed, students will find it difficult to follow and connect the missing semantic relationships. Moreover, we have assumed that students have the necessary language ability to discern the implicit semantic relationship, for example, understanding that simple and fractional distillation are hyponyms of distillation based on the English grammar. This might not be the case for many students whose native language is not the official medium of instruction.

Critical semantic connection requires us to "mind the gap" (borrowing the popular phrase from the London Underground) in the thematic pattern that typically occurs in classroom talk. This can be done before and during instruction. Before teaching a concept, it will be useful for teachers to examine a key passage from a curricular material that will be read by the students (e.g., textbook, lesson note, webpage) and discern the key semantic relationships in that passage. Drawing a thematic pattern based on the approach I have outlined earlier can be valuable as it provides an explicit visual aid to see the underlying pattern of semantic relationships. Specific attention should also be directed at the taxonomic and logical relations in the text passage, which tend to be subtle and not explicitly stated. A familiarity with key semantic relationships is beneficial because it reduces the chance that the critical links will be left out during instruction. For teachers without a strong background in the science discipline they are teaching, it is even more crucial to know what the critical semantic connections are.

During instruction, it is always a good practice to clarify the semantic relationships made by the students at every stage before going further into the content development. Take a look at the following example where a teacher asked a pair of students what they had discussed earlier based on their observation of a transverse wave motion in a water ripple tank (turn 1), and later clarified with them what they meant by "particles" (turn 3):

EXCERPT 4.3

#	Speaker	Utterance	Semantic Relationship
1	Teacher	Kate? What did you all discuss?	
2	Kate	The particles don't move sideways, they move up and down	particles − [material] − move − [manner] − up/down, (not) sideways
3	Teacher	When you refer to particles, what particles are you talking about?	
4	Kate	The particles in the water	Particles − [location] − water
5	Teacher	**Are you talking about water molecules?**	Particles − [hyponym] − molecules − [hyponym] − water
6	Kate	(*Nods her head*)	molecules
7	Teacher	So you are saying water molecules. I mean, particle is nothing wrong. **I just want to be more specific.**	
8	Teacher	When you say "particles in the water," it can mean other particles inside the water. Like some impurities moving with the water molecules	Particles − [location] − water Particles − [hyponym] − impurities

At the beginning of this conversation, Kate provided an appropriate reply to the teacher's question in terms of describing the particles in a wave motion as moving up and down rather than traveling along the direction of the wave motion. However, instead of accepting this answer and further developing the idea, the teacher took a step back to clarify with Kate by asking her to be more specific. Kate's response in turn 4 was ambiguous, which prompted the teacher to "re-voice" (see last chapter) her response to "water molecules." Semantically, these two phrases − "particles in the water" and "water molecules" have very different meanings. The first phrase is a circumstantial relation with particles as located in the water while the second phrase is a taxonomic relation of water molecule as a type of molecule (H_2O), which is itself a type of particle. The teacher made a remark in turns 7–8 to make explicit their differing semantic relationships. He first mentioned that "water molecules" was more specific than "particles" in turn 7. He then contrasted this relationship to another possible hyponym relationship – impurities as a type of particles – that could be interpreted based on the phrase "particles in the water." By making these connections explicit, the teacher avoided a potential semantic gap that was not made clear to the students.

As illustrated in this example, there are various ways teachers can prompt students to take note of the missing links in the thematic pattern. A simple approach is to make use of follow-up questions in an IRF interaction as discussed in Chapter 3. For example, asking probing questions like, "What particles are you talking about?" (turn 3), or more specifically, "*What kind* of particles?" and "Give me *an example of* a particle." These questions are suitable as they prompt students to think about the hyponym relationships, which tend to be implicit in the thematic pattern. A paraphrase is also suitable as it corrects a student's ambiguous expression into one that articulates the required semantic relationship. To complement the paraphrase, it is valuable to give a commentary over how the student's expression differs from the rephrased expression in terms of their semantic relationships. For example, the teacher compared the two expressions "particles in the water" and "water molecules," and used a form of metadiscourse (e.g., "when you say") to make a commentary of their differing semantic relationships. Such commentary involves the use of metadiscourse to talk about the talk itself. It is a useful discursive resource that we will cover in more detail in Chapter 5.

Unpacking Abstraction

Science is often regarded as a difficult discipline due to its abstract nature. As I have explained earlier, the reason why many scientific ideas are abstract is because of the language that is needed to formulate those ideas. One characteristic feature commonly found in scientific language is "nominalization" (Fang, 2005), which is a grammatical process that converts verbs or adjectives into nouns (Halliday, 1993b). Nominalization transforms common experiential processes (expressed by verbs and adjectives) into new entities or technical terms (as nouns), which can then be further described, reasoned, and synthesized with additional clauses and sentences, thus packing more information into the term. Through successive addition of meaning into the technical term, first achieved through nominalization, this is what gives certain technical words or concepts its abstract nature. Nominalization is not a stylistic choice where scientists deliberately make their language more abstract. Instead, nominalization is an indispensable factor that led to the development of sophisticated thinking and advance knowledge in science (Halliday, 1993a).

The term "distillation" in the last example is a good illustration of nominalization. Distillation is a noun that comes from the verb "distill" – a material process synonymous to "extract" that involves physical objects and actions. After it is nominalized, "distillation" becomes an entity that can subsequently be talked about and reasoned at a higher level of abstraction. For example, after nominalization the entity can be classified (simple vs. fractional distillation), given an attribute (simple distillation is easy), and argued (why use fractional distillation?). In the process of nominalization, the human actions and physical objects that are joined to the material process verb are detached and become non-essential; for example, in the statement, "We *distill* water from seawater using a flask," the pronoun "we" and object "flask" are removed when the statement becomes normalized as "distillation of seawater." Subsequently, the term "distillation" can now be joined with other words to turn into a noun phrase, such as "distillation flow rate," which is another way of packing more meaning into a few words. Finally, this noun phrase can be further abstracted into a scientific symbol (i.e., D) and used to form relationships with other symbols in a mathematical formula. Thus, through every step in the transformation of language, this is how we turn from an experiential world of things and actions into an abstract world of mathematical and taxonomic relations, with mathematical relations more commonly found in the physical sciences and taxonomic relations being more common in the biological sciences.

As nominalization and noun phrase condense a lot of information into technical terms and make it difficult for students to understand, the remedy is to unpack the semantic relationships that were condensed in those terms. This requires more than recalling the standard definition associated with the terms, which many teachers tend to do in the classrooms. What is more important is a word-by-word expansion of every term and helping students make the semantic connections from those words. As an example, the following conversation centered on the phrase "series of compression and rarefaction," which is an extended noun phrase commonly used to explain how a sound wave is produced:

EXCERPT 4.4

#	Speaker	Utterance	Semantic Relationship
1	Teacher	How does sound wave travel through air?	Sound − [material] − travel − [manner] − through air
2	David	Sound travels through a series of compression and rarefaction	Sound − [material] − travel − [manner] − through series of compression and rarefaction
3	Teacher	Wow. Where did you get that from?	
4	David	I read the textbook	

(Continued)

EXCERPT 4.4 (Cont).

#	Speaker	Utterance	Semantic Relationship
5	Teacher	Okay. So you copied from the text-book. What do you think it means?	
6	David	Erh . . .	
7	Teacher	Let's try to unpack every word. First, what do you understand by series?	
8	David	Uh. (*Points at the diagram from the textbook.*) This is the compression, this is the rarefaction. This is compression, this rarefaction. And this is compression	(Diagram) – [identifying] – compression (Diagram) – [identifying] – rarefaction
9	Teacher	Okay. So there is many of them in a series right?	
10	Teacher	What about compression? Lina?	
11	Lina	Compression is the high pressure area	Compression – [identifying] – high pressure area
12	Teacher	Okay. But it's still very abstract and doesn't say anything about how sound travels? . . . What about the word compression itself? Where do you get this word from?	
13	Lina	Compress?	Air particles – [material] – compress
14	Teacher	Compress. So what is compressed?	
15	Lina	Air particles	
16	Teacher	Yes, okay. Air particles to particles are being compressed together. Then how do you connect the air particles to what you said earlier about high pressure area? . . .	
17	Lina	More particles packed together	Particles – [material] – packed – [manner] – together
18	Teacher	That's right. And what makes them packed together? We discussed this before many times before last few weeks . . . What's that?	[reason] – air particles – [material] – collide
19	Class	Collision	
20	Teacher	Yes. It's the collision of the air particles, remember? So how do you explain that? . . .	

In response to the main question posed by the teacher, David's reply in turn 2 was a direct appropriation from the textbook – a kind of manifest intertextuality (Bakhtin, 1986). This statement from the book, in terms of the sentence construction, is not difficult to understand; it essentially contains a material process relationship (i.e., "sound travels through …"). However, the key information is condensed in the circumstantial relation within the phrase "a series of compression and rarefaction." This phrase is an extended noun phrase that was built up through a nominalization of two words – compression and rarefaction – and further combined with "a series of."

From David's reply, the teacher's instinctive follow-up questioning was to ask him where he heard the statement (turn 3) and what it meant (turn 5). He then proceeded to guide them in unpacking every word in the phrase, with "series" (turns 7–9), "compression" (turn 10 onwards), and eventually to "rarefaction." Particularly noteworthy was the way the teacher unpacked the student's phrase "compression is the high pressure area" (turn 11). He made it clear that the phrase is too abstract and does not address the key question of how sound travels. He then asked about origin of the word "compression." This action helped the students to "reverse nominalize" the abstract noun back to its material verb form – "compress." After this change, the dominant semantic relationships began to shift from abstract relational process: compression – [identifying] – high pressure in turn 11, to more concrete material processes: air particles – [material] – compress (turn 13), packed (turn 17), and collide (turns 18–20). We also see how by reverse nominalizing, this allowed the teacher to continue asking follow-up questions (e.g., probe, extend) and keep the conversation going through this IRF interaction.

Jumbled Sequencing

While *critical semantic connection* and *unpacking abstraction* focus predominantly on transitivity and taxonomic relations, the next approach, *jumbled sequencing* (adapted from Grant, 2002), centers on supporting students in making logical relations in the thematic pattern. Logical relations, particularly causal and temporal sequences, form the main logic of a scientific explanation (Unsworth, 2001a). Logical relations are formed by joining successive clauses together through conjunctions, such as *because, thus,* and *therefore* for causal relation, and *first, then,* and *finally* for temporal relation. When students learn scientific explanation, they are seldom given an opportunity to construct the logic of the explanation by piecing together the logical connections among the numerous clauses found in an explanation. Furthermore, most scientific explanations are presented to them in the correct sequence linearly from the beginning to the end of the explanation.

In helping students learn a scientific explanation, one useful strategy is to disrupt the logical sequence in that explanation by jumbling the sequences that are presented to the students. This strategy called *jumbled sequencing* involves printing the sequences of a written explanation on numerous paper strips and getting the students to rearrange them in a logical deductive order. The conjunctions in the explanation should be removed so as to allow students to rearrange the explanation by thinking about the underlying logical sequence rather than examining the grammatical relationships based on the conjunctions. After the students had completed the sequence, they could then write down the appropriate conjunctions that join the sequences they had arranged.

In the following example, the teacher used jumbled sequencing to aid her students in explaining why potassium chloride can conduct electricity in molten state but not in solid state. After the students had completed this activity, the teacher nominated a pair of students to show their order of sequence to the class through a projector. As shown in the order of the strips from top to bottom in Figure 4.4, some of their sequences were incorrect (i.e., not deductive). The teacher thus used the students' sequences to engage in a discussion with the whole class on the correct sequencing. (The numbering next to the strips in Figure 4.4 were written after the discussion).

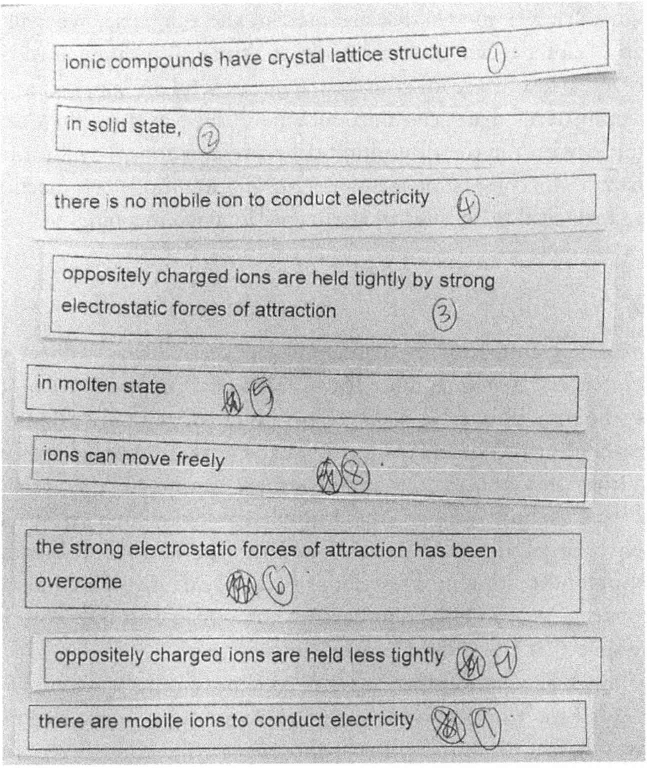

FIGURE 4.4 A jumbled sequencing activity

Other Strategies Related to Non-Verbal Modes

The strategies introduced above are useful when the instructional focus centers on spoken or written words. However, science meaning-making is seldom achieved through a verbal mode alone. It is always multimodal as scientific concepts are constructed through a combination of verbal, visual, gestural, and material-operational modes (Lemke, 1998). In the above examples, non-verbal modes were an important part of the meaning-making even though they were not highlighted. For instance, the discussion on the "series of compression and rarefaction" in Excerpt 4.4 was mediated by a diagram in the textbook showing the sound compression and rarefaction. The jumbled sequencing strategy also tapped on a visual mode of organizing and ordering written clauses (from top to bottom).

As we explore the use of non-verbal modes in Chapter 8, we find that the ideas around semantic relationship and thematic pattern are equally applicable in all semiotic modes besides the verbal mode. The reason why I have focused mainly on verbal mode in this chapter is because it is easier to understand the idea of thematic pattern based on spoken and written words before I expand the idea to other semiotic modes. As thematic pattern is still the key to understanding the content of science through whatever modes that are used in the discourse, we will revisit this idea again in Chapter 8 when we focus on multimodal meaning-making. In addition, there are a few more discourse strategies useful in supporting students to form thematic patterns that I can introduce here. However, as these strategies are better illustrated with the use of multimodal representations, I will introduce them later in Chapter 8. These strategies are critical multimodal connection, explicit comparison, analogical reasoning, and concept-language mapping.

Summary

For many teachers and students, science is a content-heavy subject with many essential "concepts" that one must learn and know in order to be considered scientifically literate (Roberts, 2007). The NGSS for instance prescribes a list of core ideas and cross-cutting concepts in the major disciplines of natural science that all students must know at various age groups. These are the curricular content that is built around the knowledge developed in science. Many educators have largely ignored the role of language and classroom discourse in the development of content knowledge, based on an erroneous assumption that concepts can exist independently from discourse, and therefore language is merely a vehicle that transmits a pre-existing concept.

As this chapter has demonstrated, what we know of as a scientific concept is essentially a label for a network of semantic relationships among words and symbols that are shared, institutionalized, and repeatedly constructed on multiple occasions and settings around the world. As such, there is an integral relationship between classroom discourse and science content knowledge. In particular, how this knowledge is formed in the classroom can be understood through the lens of thematic pattern. This understanding facilitates a more explicit way of planning and using discourse strategies that can better support students in making the critical semantic relationships as they learn scientific concepts.

Interaction pattern (discussed in Chapter 3) and thematic pattern (this chapter) form an inseparable basic unit in classroom discourse that explains how the content of science is talked into being through the interaction among the participants. One cannot take place without the other. However, these two patterns are incomplete in accounting for another important aspect of classroom discourse; that is, how a teacher's actions and performance shape a science lesson into a kind of "narrative" that is both organized and interpersonal. To examine this aspect of classroom discourse, we need to look into the role of metadiscourse, which is a kind of commentary made by people about their own talk. This will be the focus of the next chapter.

Notes

1 For simplicity reasons, this chapter will focus on the linguistic mode of representation based on spoken and written words. In Chapters 7 and 8, a more comprehensive account that incorporates all modes of representation (e.g., diagrams, gestures, mathematical symbols) will be presented.

2 In Halliday's framework, each of the attributive and identifying processes have three main types: intensive, circumstantial, and possessive. This gives rise to six categories of relational process. In this book, I do not see the need to make this distinction.

5

USING DISCOURSE TO

Organize and Evaluate Scientific Narratives

How does an engaging teacher talk science with students in a way that is constructive, organized, and coherent to them? Many teachers may undoubtedly think of using questioning and other interaction strategies to engage the students. This was our focus in Chapter 3. But besides interaction discourse strategy, there is another way which tends to be overlooked by most teachers, and this involves the use of metadiscourse (or meta-talk) to talk about the talk itself. When experienced teachers introduce a new concept or explain a phenomenon, they do not simply deliver the content matter by just telling them or asking questions about it. Rather, they tend to also organize the content into a rhetorical and relational "narrative" and provide an ongoing commentary on the unfolding narrative in order to make it more accessible for students to follow (Scott, 1998). This commentary on the narrative, which runs alongside the development of the content matter, is enabled through the use of metadiscourse. Some common examples of metadiscourse include connecting to past conversation (e.g., "as I was saying just now"), focusing on current topic (e.g., "what are we talking about now?"), marking significance (e.g., "listen, this is important"), and expressing affect (e.g., "isn't this cool?").

In Chapters 3 and 4, we saw how scientific knowledge is constructed through the interaction and thematic patterns made jointly by the teachers and students. Through successive interactional exchanges (e.g., IRE, IRF) and connections among multiple semantic relationships, this is how every piece of knowledge is incrementally assembled within the social plane of the classroom (Vygotsky, 1986). This way of conceptualizing and analyzing classroom discourse is useful in describing and analyzing how the teaching of science takes place in the classroom. However, it is incomplete as it does not consider the extended temporal continuity in the classroom as well as the teacher's view in putting a lesson together. This is where the term "narrative" as coined by Scott (1998) is suitable in describing the deliberate performance of a teacher in orchestrating a science lesson. The word narrative does not imply that the scientific knowledge in the lesson is fictional or imaginary like a story. However, like a good story, a science lesson needs to be carefully crafted with the audience (i.e., students) in mind such that it is organized and accessible. Moreover, a science lesson does not just contain facts, but like any other story, it also conveys subtly or overtly the author's affective and ideological stances toward the story.

There are two aspects (and consequently patterns) of narrative in science lessons. The first is a cohesive aspect in terms of how the narrative is organized and integrated with numerous discursive elements. This is somewhat related to Bakhtin's notion of intertextuality and Halliday's textual metafunction of language as discussed in Chapter 2. The second is an interpersonal aspect in terms of a particular stance toward the science content, and this is related to Bakhtin's notion of voice (ideological stance) and Halliday's interpersonal metafunction of language (also in Chapter 2). Both of these aspects in a lesson narrative are very seldom explicitly mentioned or discussed in classroom discourse. However, they are always present, and more importantly for us, they are revealed occasionally when the participants are talking about their own discourse (i.e., metadiscourse). Such metadiscourse can come in the form of a direct commentary on what is said, or more subtly expressed through some words within an utterance, for example, pronouns (e.g., we, I), conjunctions (e.g., just now, later), and modal adjuncts (e.g., probably, logically).

In this chapter, I will first elaborate on the narrative patterns in classroom discourse and how they are connected to metadiscourse. This will be followed by an extended discussion of the various types of metadiscourse strategies and how science teachers use them to organize and evaluate the science content they are teaching. I will then discuss how a conscious and judicious use of metadiscourse can make science teaching more comprehensible, organized, relevant, and engaging to students.

Narrative Patterns of Science Lesson

To help us understand narrative patterns and how it is tied to metadiscourse, let us re-examine an example that had been discussed previously in Chapter 4 to illustrate thematic pattern (see Excerpt 4.2). In Excerpt 5.1 below, the same transcript is reproduced with some changes. First, I have truncated some parts of the dialogue that focus on the content development with few instances of metadiscourse (lines 2–12 and lines 15–18). Second, I have replaced the last column on *semantic relationship* with *metadiscourse* in this excerpt. The specific types of metadiscourse in this column (e.g., sequencer, past conversation) will be elaborated in a later section.

EXCERPT 5.1

#	Speaker	Utterance	Metadiscourse
1	Teacher	**Now next**. Why do I choose to use fractional distillation rather than simple distillation?	Sequencer
2–12	Class	*(content development of simple distillation)*	
13	Teacher	Okay, **so that is one way** that you use simple distillation	Past conversation
14	Teacher	**But** why is it that we use fractional distillation here? What's the main reason?	Sequencer
15–18	Class	*(content development of fractional distillation)*	
19	Teacher	Alright, **so first of all**, liquid air is a mixture of liquids, alright, having similar or, or, or, close boiling points	Sequencer
20	Teacher	Alright, **so therefore** I use fractional distillation	Sequencer
21	Teacher	Okay, **and for** simple distillation, we use it for soluble solid-liquid mixtures	Topicalizer
22	Teacher	Alright, for example, **you might want to write it in**	Importance
23	Teacher	**just now** you're telling me about seawater to become salt plus distilled water	Past conversation
24	Teacher	So **that is one example that you shared with me**	Past conversation

In terms of the content development, this whole-class discussion has three thematic segments. The first part was an expansion on simple distillation from line 1 to 13, while fractional distillation was expanded subsequently from line 14 to 18. We saw in Chapter 4 that the semantic relationships for simple and fractional distillation were very similar (see thematic pattern in Figure 4.3). The last part from line 19 to 24 was a summary of the discussion in this excerpt, as evident from the repetition of the key semantic relationships and the monologic talk by the teacher.

Each of the above-mentioned thematic segments were joined together into a cohesive unit through a category of metadiscourse called *text connective*. The function of text connective is to maintain continuity and coherence in any conversation. Three types of text connective were seen in this excerpt, namely past conversation, sequencer, and topicalizer. For instance, in line 13, the phrase "so that is one way" functioned as a text connective to sum up and signal the end of the discussion for simple distillation. In particular, the deictic expression "that" referred to whatever that had been said from line 2–12, thus it serves as a text connective to a *past conversation*. In other words, "that is one way" in line 13 is a talk about the talk that has transpired earlier, and this is why it is a metadiscourse.

As line 13 signaled the end of discussion on simple distillation, the shift to a new subtopic (fractional distillation) was signaled in line 14. This was done with the conjunction "but," which highlighted a contrasting sequence to what had been discussed earlier. Thus, this is a *sequencer* text connective. Then in line 19, another important transition occurred and this is marked by the sequencer in "so first of all" as well as a high occurrence of several other text connectives from line 20 onwards (i.e., topicalizer, past conversation). Collectively, these text connectives worked to signal the "talk about the summary" to the students and they linked the various thematic segments that had been discussed into a summative and cohesive narrative.

Excerpt 5.1 shows a brief example of the narrative pattern of opening, closing, framing, sequencing, announcing, and summarizing within a class discussion led by a teacher. This narrative pattern is commonly seen in the introduction, transition, and conclusion stages of any thematically-bounded activity structure. Such activity structure can be a short discussion (typically a few minutes) on an explanation or question similar to what we saw in Excerpt 5.1. It can also be a larger activity (typically 30 minutes or more), such as a full lesson, comprising typically a warm-up, concept development, and lesson closure type of structure. Similar activity structure is also found in written texts in terms of its organization at the level of paragraph, section, and chapter. For instance, notice my usage of metadiscourse in the first four paragraphs in this chapter where I talked about previous chapters as well as what I am going to say in this chapter. Regardless of any activity structure, metadiscourse is indispensable to the organization of the discourse. Specifically for classroom discourse, metadiscourse provides a crucial way for teachers to control the flow of activity (e.g., discussion, lesson) by signaling boundaries and making connections between the various thematic segments.

Beside this organizational function of metadiscourse that addresses the cohesive aspect of discourse, there is another evaluative function that corresponds to the interpersonal aspect of discourse. This interpersonal aspect is not prominent in Excerpt 5.1, except for one occurrence in line 22 when the teacher asked the students to "write it in." This is an example of a metadiscourse called *attitude marker* which signals a particular attitude toward the talk they had or will be discussed. In this instance, the attitude is according to the teacher "important." Such attitude markers and other types of evaluative metadiscourse are interspersed in various parts of classroom discourse, and they indicate to the participants a kind of evaluative position or stance toward the content as aligned with the authors' attitudes or values.

Other than this attitude marker to importance, what is interesting about the lack of other evaluative metadiscourse in Excerpt 5.1 is itself indicative of a common narrative pattern when teachers talk about science in the classroom. Notably, it portrays the view that scientific knowledge is factual, objective, and self-evident. This indicates an implicit and subtle stance toward the content according to the teacher's attitudes and values. We will see in later episodes how other evaluative metadiscourse can give rise to a different stance toward scientific knowledge.

The Role of Metadiscourse: A Commentary on the Unfolding Narrative

According to linguists, metadiscourse is a type of discursive resource used to organize text (both oral and written) as well as signal the author's stance toward the content of the text (Hyland, 2015). It is a kind of non-propositional commentary made in the course of speaking or writing. As a commentary, metadiscourse does not expand the propositional content in the text. However, it has a range of cohesive and interpersonal features to help an audience connect, interpret, organize, and evaluate the content in a way that is preferred by the author or aligned with the social conventions and values of the discourse community (Vande Kopple, 2012). Metadiscourse is almost always present in most texts, even when the speaker or writer is not consciously aware of its existence and use.

There are two general types of metadiscourse – organizational and evaluative, which corresponds to both the cohesive and interpersonal features of discourse (Schiffrin, 1980). Organizational metadiscourse assists in the rhetorical organization of propositional content by signposting past and future references, signaling topic shifts, building sequences, and connecting to external ideas. At the same time, evaluative metadiscourse accentuates aspects of the propositional content by injecting the author's attitudes and stances toward the content with the use of comments, modifiers, and heteroglossic projections. Therefore, in spoken and written discourse, there are always three components happening – propositional content, evaluative metadiscourse, and organizational metadiscourse. This classification of metadiscourse is inspired from and aligned with Halliday's (1994) view in SFL theory that all language use expresses three type of meanings simultaneously: ideational, interpersonal, and textual (see Chapter 2). In the context of science classroom discourse, propositional content refers to the ideational meaning of science (i.e., scientific knowledge defined within a curriculum), while evaluative and organizational metadiscourse consist of linguistic resources from the interpersonal and textual functions of language.

The research on metadiscourse first originated from language education and applied linguistics, particularly in the area of academic discourse, written communication, and English for language learners (e.g., Jalilifar & Alipour, 2007; Lee & Subtirelu, 2015; Vande Kopple, 2012). However, there are few systematic studies that examined metadiscourse in classroom talk, largely because it is more complicated. Furthermore, research in metadiscourse within science education is almost non-existent. In recent years, I did a pioneering study to systematically examine and document how several science teachers used metadiscourse to construct scientific knowledge with their students (see Tang, 2017). The result of this study was a typology of metadiscourse that is commonly found in science classroom discourse, which will be introduced next.

Metadiscourse Strategies

Metadiscourse is technically a form of discourse strategy according to our definition of discourse strategies in Chapter 1, which is "methods that people employ to understand each other in a conversation to achieve certain purpose" (Gumperz, 1982). By using metadiscourse to talk about our own discourse, metadiscourse creates and accentuates the underlying narrative patterns of our conversation, thereby assisting our audience to connect and relate to our message in a more accessible way. The key issue here is that most teachers are not aware of the existence of metadiscourse and thus miss the opportunity to harness and control this discursive resource for their classroom discourse. In this section, we first explore several types of metadiscourse that are commonly used, consciously or unconsciously, in the science classrooms based on my empirical study (Tang, 2017). This will be followed by a discussion of how metadiscourse can be more consciously developed and deliberately used by teachers to explicitly support students in following the narrative of a science lesson.

Types of Metadiscourse

The use of metadiscourse can generally be classified into six categories with different functions, as shown in Table 5.1. These categories are text connective, knowledge connective, and activity connective for organizational metadiscourse, and attitude marker, epistemology marker, and interpret marker for evaluative metadiscourse.

TABLE 5.1 Classification of metadiscourse in classroom discourse

Type	Category	Function
Organizational	Text connective	Connect one part of a conversation to another; includes *past conversation, future conversation, sequencer, topicalizer*
	Knowledge connective	Connect one's knowledge to current conversation; includes *prior knowledge, apply knowledge*
	Activity connective	Connect one's activity to current conversation; includes *previous activity, ongoing activity, external activity*
Evaluative	Attitude marker	Signal one's attitude toward propositional content; includes *affect, importance, challenging*
	Epistemology marker	Reflect one's stance toward evidential status of propositional content; includes *sensory experience, logic, scientists' work, postulate*
	Interpretive marker	Direct others to grasp or construct the appropriate interpretation aligned with a particular stance, attitude or point of view; includes *modifier, paraphrase, projection*

Text Connective

One of the most common types of metadiscourse is *text connective*, which links one part of a conversation (or text in general) to another so as to forge continuity and coherence. Thus, they are used as a discursive resource to maintain the historicity of texts, or intertextuality (see Chapter 2). Text connective includes referencing something that was mentioned earlier as well as creating expectation to a later conversation. Text connectives provide an important resource for teachers to relate an idea to a prior or future discussion, as seen from the following two familiar examples:

EXCERPT 5.2

#	Speaker	Utterance	Metadiscourse
1	Teacher	Can anyone remember what was the main thing **we talked about in the last lesson**? Josh?	Past conversation
2	Josh	Pressure	
3	Teacher	Yes, pressure. Or more exact, **we talked about** water pressure	Past conversation
4	Teacher	In **today's lesson,** we are still in the topic of pressure, but **we will learn about** air pressure...	Future conversation

EXCERPT 5.3

#	Speaker	Utterance	Metadiscourse
1	Teacher	Okay. This second question "How do they work together?" **You already said** there are so many things. And **just now we already mentioned** the function of each part. The blood, the heart, and the blood vessel.	Past conversation
2	Teacher	So now, if **you are going to describe** how do they work together, **I want you to use,**	Future conversation
3	Teacher	**just now you said** the functions, the parts and some of the words here.	Past conversation

Excerpt 5.2 shows a very common exchange seen in a typical lesson introduction where the aim was to make a connection from the main topic learned in a previous lesson to the current lesson. The text connectives to past conversation in lines 1 and 3 serve as a recap while those that signpost to future conversation in line 4 serve as an advanced preview to the key concepts of the day's lesson. In Excerpt 5.3, the text connectives were not used for a lesson introduction, but were used instead to connect immediate ideas that were discussed in the middle of a lesson. In this case, the students had previously mentioned the parts and

functions of different human organs (lines 1 and 3). Thus, the teacher integrated those past conversations together in order to facilitate the students to make the connection to the next question, which is "how do they work together?"

Mercer (2000, p. 52) recommends a variety of techniques for teachers in "building the future from the past." Three of these techniques – recap, elicitation, and exhortation – make explicit connections between the past and present in order to maintain a temporal continuity within the teaching narrative. These techniques always involve metadiscourse in the form of text connectives, which can be found in the conversation. Typically, text connectives tend to involve pronouns and verbal processes (e.g., we talked about, you said, we are going to discuss), temporal adjectives and adverbs (e.g., just now, yesterday, today, right now, later), and some words like "remember," "recall," or "think back."

Besides signposting to past and future, text connectives are also useful in organizing a conversation into different "conversational topics," labeling them, and subsequently making reference to them. This is the use of *topicalizer*, which Williams (1981, p. 50) defines as words that "focus attention on a particular phrase as the main topic of a sentence, paragraph, or whole section." Excerpt 5.4 shows an example where topicalizer was used in lines 1, 7, and 11:

EXCERPT 5.4

#	Speaker	Utterance	Metadiscourse
1	Teacher	**For scenario 1a, we were talking about** 2 skaters pushing against each other.	Topicalizer Past conversation
2	Teacher	So **we saw** Thomas and Alan pushing one another. Applying the force on one another. So **what did we observe** over here? Sorry, you were saying?	Previous activity
3	Aaron	Move away from each other. . .	
4	Marco	Both then move back	
5	Chin	At the same speed. . .	
6	Teacher	I think **just now** Sam made a good point that actually the distance traveled by both are almost similar	Past conversation
7	Teacher	**And then when we moved on to 1b (*writes 1b*)**, where only Alan was pushing Thomas.	Topicalizer
8	Teacher	When Alan was pushing Thomas right, **what was the observation?**	Previous activity
9	Aaron	Both of them moved backwards. . .	
10	Teacher	But, **isn't it weird?** Because, in this case, both exert a force each (*writes "both exert a force" under 1a*). But in this case right, only A is pushing B. (*writes "only A is pushing B" under 1b*). So how come B also start to move? . . .	Affect
11	Teacher	Okay. **Later we will see for scenario 2**, which we will have a spring balance being pulled. . .	Future conversation + Topicalizer

The conversational topics in Excerpt 5.4 were three contrasting scenarios (called scenario 1a, scenario 1b, scenario 2), each referencing to an event that had occurred prior to the conversation. The topicalizers, such as "for scenario 1a" (line 1) and "moved on to 1b" (line 7), were used to direct students' attention to these topics and their referential content that followed. Very often, topicalizers are accompanied by text connectives to past and future conversation (e.g., were talking about, later). However, topicalizers play an additional role in condensing previously discussed information into short labels and then juxtaposing these labels for quick contrast and comparison. Once a topic of discussion has been established through the use of topicalizers, they can then become a shared linguistic resource to be used by the participants for making references to their own talk.

Another type of text connective is *sequencer*, which is typically marked by conjunctions[1] (e.g., first, then, so, but) and words with temporal meaning (e.g., start, later, step). We have seen some examples of sequencer earlier in Excerpt 5.1. Sequencer can also be made through visual cues to highlight temporal or causal sequence of events, steps, or development of ideas. An example could be the use of a visual flowchart such as Figure 5.1 to mark the boundary and sequence of ideas. Even the use of presentation slides like PowerPoint to segment chucks of ideas into individual slides to be presented temporally is a form of sequencer text connective (Stoner, 2007).

Activity Connective

Activity connective is slightly different from text connective in that it does not make linkage to any specific information or content, but rather directs attention or connection to a *previous* or *ongoing activity* accompanying the talk or an *external activity* outside the talk. In Excerpt 5.4, we saw two instances in line 2 and 8 where the teacher made direct reference to an experiential activity where two students (Thomas and Alan) were pushing each other while standing on

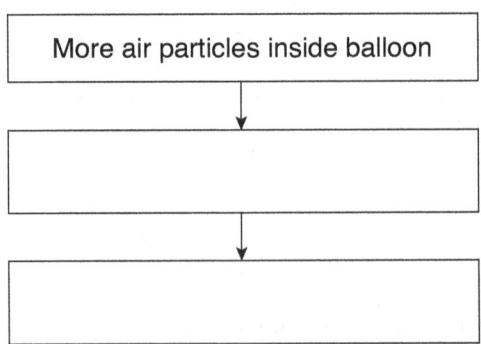

FIGURE 5.1 An example of sequencer text connective in a visual flowchart

a skateboard. The activity connections through phrases like "we saw," "we observe," and "what is the observation" made the connection between the talk and the activity.

Activity connective is not limited to connecting to an experiential activity. It also includes marking the actions and work processes involved in an ongoing task. The following excerpt shows a discussion between two students as they were exploring a computer simulation to solve a particular task. An activity connective was used by Krishna in line 1 to highlight the action she was going to make and Leala in line 8 when she pondered upon the current task they were doing:

EXCERPT 5.5

#	Speaker	Utterance	Metadiscourse
1	Krishna	**I want to click this. Then we can play with** the battery thing in parallel.	Ongoing activity
2	Leala	(*observing the simulation*) **Wow. So cool**...	Affect
3	Krishna	**I think** series battery will be brighter	Postulate
4	Leala	Wait. Parallel batteries. **Are you saying** the parallel bulbs or batteries?	Paraphrase
5	Krishna	Series battery will make the bulbs brighter. Parallel will...	
6	Leala	Parallel arrangement of the bulb or of the battery. **Which one you saying?**	Paraphrase
7	Krishna	Yah. I'm **confused**...	Challenging
8	Leala	So what are **we doing now?**	Ongoing activity

Excerpt 5.5 provides an example that shows metadiscourse is not exclusive to teacher talk or a discussion facilitated by the teacher. As a discursive resource for human communication, they are also found in student dialogue, albeit with different emphasis compared to most teachers' use of metadiscourse. Students tend to use less organizational metadiscourse as they seldom have the goal or experience to make the propositional content more accessible to their peers. By contrast, student dialogue tends to contain more evaluative metadiscourse, such as affect, postulate, paraphrase, and challenging as seen in Excerpt 5.5.

Another type of activity connective is to make linkage to an external activity that is not occurring in the classroom but is nonetheless relevant to the current talk. Such metadiscourse is used to draw upon resources outside the classroom to contextualize the meanings made in the classroom. This includes the recounting of specific or general events from the students' experiences (Pappas, Varelas, Barry, & Rife, 2004). For example, in teaching topics with many real-world examples and applications, teachers frequently make reference to familiar activities, such as food, sports, or everyday technology, as seen in this example from a teacher:

> **Remember I told you about the marathon?** For example, if I ask Yoon, a younger kid to go and run against an older kid. Who do you think will win?

Knowledge Connective

The last category of organizational metadiscourse is knowledge connective, which serves to make links between the students' knowledge and the content of the talk. Such linkages are useful for teachers to relate to students' prior knowledge or get students to apply knowledge to the current conversation. An example is shown here:

> **From your knowledge from year 7 and 8**, when you were learning fractional distillation, alright, can you try to predict with a reason, which of these gas will be distilled out first? **I'm trying to link back to what you've learnt before.**

In many ways, knowledge connective is very similar to text connective in terms of connecting to a prior event. In fact, informed by Vygotsky's (1986) theory of internalization, one can argue that a knowledge connective is really a text connective to a past conversation from a long time ago. However, for practical reasons, it is useful to distinguish knowledge connective as a reference to information that cannot be traced back to a specific text or conversation, usually because the event has transpired by several weeks, months, or years. In addition, knowledge connective is also dependent on the audience (i.e., students' knowledge), while text connective, being traceable and located in the text, is not so dependent on the audience.

Attitude Marker

While text, activity, and knowledge connectives function as organizational metadiscourse, *attitude marker* is a type of evaluative metadiscourse. Its primary role is to signal one's attitudes toward the propositional content being discussed. The expression of *affect* is a type of attitude marker as it is a commentary of the unfolding narrative and signals a range of emotional expressions toward it, including surprise, excitement, intrigue, amusement, or boredom. We have seen some examples of affect in Excerpts 5.4 and 5.5.

Another type of attitude marker is *importance*, which is often used by teachers to evaluate some parts of the content that are being taught. It signals the significance of understanding or recalling specific content matter. An example was shown in Excerpt 5.1, when a teacher gave a specific instruction to "write it in" the talk that has just been discussed. Another typical example would be making remarks such as "this is important" or "you need to take note of this."

A closely related attitude marker is *challenging*, which revolves around comments about the difficulty of the propositional content being discussed. This is used as a way of mentally preparing the students for the forthcoming challenge when discussing a new instructional task or topic; as shown in these examples: "We are going to the *difficult* part, which is crafting explanation using kinetic model of matter" and "Some of you are facing *difficulties* with this chemical equation right?" This attitude marker can also be signaled by students when they say "I don't understand" or "I'm confused," as we saw in Excerpt 5.4.

Epistemology Marker

Epistemology marker signals a person's epistemological stance toward the evidential status of the propositional content, in terms of how do we know the information being communicated is true? This category of evaluative metadiscourse is particularly useful in signaling a number of possible evidential bases, such as sensory experience, logic, scientists' work, and postulate.

Sensory experiences are evidential stances based on observations of an object or phenomenon seen from a demonstration, experiment, or video. For instance, in a demonstration to show the effect of height on water pressure, a teacher made three holes of varying height on a plastic bottle and poured water into it. As water began to spurt out from the three holes, he made a running commentary of the phenomenon, which revealed his epistemological stance:

EXCERPT 5.6

#	Speaker	Utterance	Metadiscourse
1	Teacher	So, **we're going see** whether point A, point B or point C has the highest pressure	Ongoing activity
2	Teacher	And **how do we see that**, right? Simply, it's by because, (*handling the apparatus*). Okay. So **how do we judge that**? Simply, if it can shoot the furthest. Okay, then it has the highest pressure...	Sensory
3	Teacher	Ah, now, when, when the cap is removed, right, **we see what happen**, okay?	Ongoing activity
4	All	[Demonstration followed by group discussion]	
5	Teacher	Okay, **as we can see right, from the experiment**, roughly how does it look like, right? I'll draw it for you the, the spurting. Okay, so, this one will **actually spurt like this** (*draws a curved line on the board*)... Alright, **so we can see**, C **actually** has the highest pressure	Sensory

For this teacher, the purpose of the demonstration was to let his students "see" that point C (lowest hole in the bottle) would have the highest pressure because "it can shoot the furthest" (line 2). The teacher also used activity connective metadiscourse to direct the students' attention to the demonstration, as seen in lines 1 and 3. But in lines 2 and 5, the teacher went beyond just telling them to observe the demonstration. He also established the evidential basis for determining which point has the highest pressure. In line 2 for instance, he remarked, "How do we *judge* that? Simply, if it can shoot the furthest." Also in line 5, he said, "this one will *actually* spurt like this." The adverb "actually" uttered twice in line 5 was particularly telling as it suggests some kind of "truth" in what was going on based on what was "seen." Thus, these commentaries to the source of evidence revealed the epistemological stance of the teacher. Such an evidential status based on "seeing is believing" is quite common whenever teachers and students comment "as you can see" when they are explaining something with a demonstration.

Another type of epistemological marker that reflects the evidential status of a statement is *logic*, usually a causal or deductive type of logic in science. This could be signaled through commentaries like, "by deduction," "a logical flow of answer," and "both methods are accepted as long as it's logical." Such remarks gave the impression that some explanations or "answers" in science are logically deduced from existing knowledge. Logic epistemological markers can also be signaled through non-verbal means like a flowchart. Besides serving as a text connective (sequencer), a flowchart can also function as an epistemological marker. We will see more examples of logic as an epistemological marker later.

Besides sensory experiences and logic, another epistemological marker can be references to scientists and their work. Such epistemological markers can range from comments such as "*scientists believe* that climate change is occurring" to alluding to more complex theory building and argumentation work undertaken by scientists. For instance, in a lesson where a teacher got his students to research on a topic and present their work to their peers, he wanted the students to defend their ideas against their peers' critique. At the end of the lesson, the teacher made the following comment to assure the students that the process of defending one's beliefs was "very normal" among scientists:

> It is perfectly normal. Because we all have certain sets of beliefs that we believe strongly in… Back then, all scientists do that, including Newton, Faraday. All of them trying to fight one another to see whose theory is right. Okay, it is very normal.

The last common epistemological marker is *postulate*, which signals information based on guesswork without a firm knowledge or other evidential basis. These are often found in phrases like, "I think" or "I'm guessing." For example, in the students' discussion in Excerpt 5.5, Krishna acknowledged her claim that "series battery will be brighter" is a postulate when she qualified it with "I think." It should be cautioned that some people tend to say "I think" frequently when they have some evidential basis other than guessing, and conversely, some people do not qualify their statements when they do not have any evidence. Thus, it is important to reinforce such use of epistemological markers among students by getting them to reflect on their evidential basis of their statements. We will discuss more on this issue in a later section.

Interpretive Marker

Finally, *interpretative markers* are phrases and comments to direct the audience to grasp or construct the appropriate interpretation of a text that is aligned with the stance, attitude, and point of view preferred by the speaker. There are three common types of interpretative marker: modifier, paraphrase, and projection.

Modifiers are used to modify the amount of force of our speech act and often involve modal verbs such as "must," "should," "may," and "can" to signal how strictly or loosely we want others to take our response or instruction. For instance, at the beginning of an explanation, a teacher remarked: "What are the important keywords that you need to make use of to phrase the explanation? We **want to** use 'force', we **want to** make use of 'particles', we **want to** …" In this remark, the metadiscourse signaled to the students a strict and obligatory use of vocabulary to be included in the construction of the explanation.

Paraphrases are used to signal another way of interpreting a text or conversation in order to help someone understand it or see it in a different way or point of view. Some common paraphrases are "Let me put it another way" or "What I'm trying to say here." Paraphrases are also used to compare different ways of saying the same thing from different sources, for example when a teacher compared his answer with the textbook's: "On page 159, your textbook will have explained this as well perhaps using **another way of phrasing**."

Projection is used to direct students toward the source of our utterances, using phrases like "So Mary is saying that …" or "You see where is she coming from?" It is used to acknowledge a direct appropriation of the "voice" of the statement that is being made (Bakhtin, 1986). Frequently, some teachers would want their students to adopt a scientist's identity and perspective in a task, as seen in this example: "I want you to imagine that **you are the scientist** … **you must tell people** about your discovery of the plant cell." Adopting another person's voice into one's own voice is a heteroglossic projection as it mixes two different voices and points of view (see Chapter 2 on heteroglossia).

Effective Use of Metadiscourse Strategies

From the examples of metadiscourse shown earlier, it is evident that metadiscourse is a useful discursive resource for facilitating classroom discourse in a number of positive ways. However, there is a wide disparity in terms of how science teachers are using metadiscourse in the classroom (Tang, 2017). Some teachers rarely use them as they focus more on the content of the lesson and less on the narrative aspect of making the lesson more accessible and relevant to students. On the other hand, some teachers may unknowingly use certain metadiscourse too often, to a point where metadiscourse loses its effectiveness and becomes a distraction to the propositional content. Take for example a simple phrase like, "listen, this is important." When used judiciously, it can serve as an attitude marker of *importance* to alert the students to take note of what is going to be discussed. However, when it is used repeatedly in every situation, it becomes just a habitual way of speaking that will most likely be ignored by students.

Hyland and Tse (2004, p. 157) argue that a good writer can, through the judicious use of metadiscourse, "transform a dry and difficult text into a coherent and reader-friendly prose and related it to a given context." In the same way, a deliberate use of metadiscourse in classroom discourse can help science teachers make the lesson content more comprehensible, organized, relevant, and engaging to learners. As such, a conscious awareness of the various types of metadiscourse and their functions, as summarized in Table 5.1, can help teachers enhance their classroom discourse. In what follows, I highlight a few discourse strategies that will make a more deliberate use of metadiscourse to explicitly support students in the learning of science.

Narrative Framing

As mentioned before, the teaching of science content involves a narrative structure that shapes the way students interpret the content cohesively and interpersonally. This narrative structure consists of many discursive stages, each with its own beginning, closing, and transitioning. The role of organizational metadiscourse, particularly text and activity connectives, can be used to assist students to follow the narrative structure in a lesson by marking the boundaries of these stages as well as linking them into a cohesive and temporal continuity. In addition, every discursive stage is not a simple and objective presentation of reality, but it also requires a particular evaluative stance in order to make sense of it. Such stance is often highlighted through the use of evaluative metadiscourse such as attitude, epistemology, and interpretative markers. This idea of organizing and interpreting narrative stages is closely related to the concept of framing introduced by Goffman (1974) to help us understand how teachers and students negotiate and structure "what is going on?" in any lesson activity.

Narrative framing is an important strategy that teachers can do more to help their students interpret not only the content, but also how the content is being shaped into a form of knowledge with a particular purpose (Reveles, Cordova, & Kelly, 2004). This is particularly useful when teachers are developing a concept, explaining a phenomenon, or solving a problem with their students. At the opening and closing of these activities, there are stages where metadiscourse is highly valuable. To use an analogy from written text, this is somewhat similar to the frequent use of metadiscourse at the introduction and conclusion of a section, chapter, or article. The use of metadiscourse in these stages is not just to introduce and summarize the key content ideas. More importantly, based on the notion of framing, they also prepare the students to connect and interpret the content from a particular stance. The following excerpt shows an example of how a teacher framed the opening of an explanation from line 1 to 8, before he proceeded with the usual content development through an IRF interaction pattern from line 9 onwards:

EXCERPT 5.7

#	Speaker	Utterance	Metadiscourse	
			Organizational	*Evaluative*
1	Teacher	**So how do we** explain pressure of a gas using kinetic model of matter?	Ongoing activity	
2	Teacher	Alright, **let's say this is a four marks O-level[1] question**.	External activity	
3	Teacher	**Let's try to phrase it together alright?**	Ongoing activity	
4	Teacher	**First**, we'll have, **we will start with** air particles inside the balloon. Okay, **let's start** with a balloon then	Sequencer	
5	Teacher	So we'll have, *let's use* bubble then (*Draws a box on the board*)	Ongoing activity	
6	Teacher	**Obviously** there is one more **chain of answer** here. **And then** one **chain of answer** . . . (*Draws 2 more boxes with arrows to show sequence*; see Figure 5.1)	Sequencer	Logic
7	Teacher	So let's see what are the **important keywords that you need to make use of**		Importance
8	Teacher	**We want to** use force, **we want to** make use of particles, **we want to** use area **because we want to** talk about pressure. **And then we** *want to* use them continuously, randomly, **and then** maybe collision . . . (*Writes these keywords on the board*)	Sequencer	Modifier
9	Teacher	Okay, so what happens when the air particles are inside the balloon? Brad?		
10	Brad	(*content development through IRFRF interaction sequence*)		

[1] The GCE O-level is an academic qualification offered by some examination boards in the United Kingdom and several commonwealth countries and it is usually taken at the age of 16 (year 10).

In much of this excerpt, the teacher had not started developing the explanation with his students. Instead, the teacher used an array of organizational and evaluative metadiscourse to frame the subsequent explanation to be developed from line 9 onwards. There are two aspects in this framing as characterized by the dual cohesive and interpersonal nature of narrative patterns.

The first aspect (cohesive) was to set the context for the subsequent discussion through the use of activity and text connectives. First, activity connectives highlighted the ongoing activities students were expected to do, including: (a) answering an explanation question (line 1), (b) phrasing a written response (line 3), and (c) using a particular visual aid (i.e., "bubble" in line 5). It also made a connection to the larger external activity of examination through the phrase, "let's say this is a four marks O-level question" in line 2. These activity connectives were used to help students gained a better awareness of what they were doing and why. Next, several sequencer text connectives were used to prepare the students to connect the subsequent discussion into a coherent explanation that was defined by a "chain of answers" (line 6) and a linear flowchart of bubbles (see Figure 5.1).

The second aspect (interpersonal) of the framing was to set the teacher's preferred evaluative stance toward the subsequent discussion through the use of attitude, epistemology, and interpretative markers. In this case, the preferred stance was coming from an examination-oriented stance of rationalizing several requirements for the explanation. First, the explanation must be logical as seen from the epistemological marker in phrases like "obviously there is one more chain of answer here" in line 6. Second, there was a strict interpretation of using several keywords as seen from the phrase "important keywords" in line 7 and several modifiers in "we want to..." in line 8. These requirements were consistent with the teacher's earlier remark about the "four marks O-level question" in line 2. Thus, the evaluative metadiscourse highlighted to the students how they should interpret the science content that would be developed in the subsequent explanation.

In this example, the teacher emphasized an examination-focused stance toward the scientific explanation. This is quite typical whenever teachers discuss test questions in most high schools around the world. However, there are also other evaluative stances depending on the nature of the ongoing and immediate tasks and discussions. For instance, when the classroom talk involves an investigation or demonstration, the evaluative stance tends to lean toward a sensory-driven inquiry that is almost contradictory to the ideology of a fact-driven examination stance. An example of this stance was seen earlier in Excerpt 5.6, which was incidentally made by the same teacher who spoke in Excerpt 5.7. Through the use of epistemology markers (as in bold below), the teacher framed the subsequent observation as the evidence to "judge" which point has the highest pressure:

So, **we're going see** whether point A, point B or point C has the highest pressure. And **how do we see that**, right? ... (*handling the apparatus*). So **how do we judge that? Simply, if it can shoot the furthest.** Okay, then it has the highest pressure.

Thus, narrative framing with metadiscourse provides a valuable resource for teachers to assist their students to connect and interpret science content according to a particular narration, ranging from a factual and logic-based stance for examination purpose to a sensory and empirical stance grounded in inquiry. As the narration of a lesson shifts across these various stances, it is important for teachers to use metadiscourse to highlight how the underlying stance has changed. Arguably, while the metadiscourse in Excerpts 5.6 and 5.7 is critical for this purpose, it may still be quite subtle for students to notice. What could be even better in these examples is if the teacher could explicitly discuss these various ideological stances and provide reasons for why they are different. This will emphasize the narrative structure even more and consequently make it easier for students to navigate the seemingly contradictory pictures between "science as a fun inquiry" and "science as an examination knowledge."

Emphasizing Science Inquiry: Heteroglossic Projection and Epistemological Marker

Evaluative metadiscourse, notably epistemological and interpretative marker, is also useful for teachers and students to talk about the nature of science inquiry. Science inquiry is a popular instructional approach that provides students with opportunities to engage in the activities and processes used by scientists to study the natural world (Bybee et al., 2006; Crawford, 2000). In supporting inquiry, there tends to be a lot of emphasis on "doing" the inquiry, notably carrying out investigation, performing experiment, and collecting data and evidence. Comparatively, there is less focus on talking about the inquiry itself. Reveles and colleagues (2004, p. 1123) argue that metadiscourse provides an important resource for teachers to orient their students to "talk about their own inquiry" in order to make an explicit connection between students' own experiences and scientists' disciplinary practices.

In a science inquiry activity, it is beneficial to ask students to assume the identity of a scientist and think like them (Reveles, Cordova, & Kelly, 2004). This is often done through the use of a heteroglossic projection (an interpretative marker) to direct the students toward a different voice or point of view and subsequently incorporate that voice into their voices (Bakhtin, 1986). This creates an opportunity to infuse science inquiry practices into the students' literacy practices, as seen in the following example:

EXCERPT 5.8

#	Speaker	Utterance	Metadiscourse	
			Organizational	Evaluative
1	Teacher	**Now I want to introduce** a new kind of science activity.	Ongoing activity	
2	Teacher	And **this activity is you are going to be a scientist** and you discovered something...	Ongoing activity	Projection
3	Teacher	This discovery that you have made is a new type of plant cell.		
4	Teacher	Congratulation! You are going to get an award.		Affect
5	Teacher	But before you can get this award, **you must tell** people about your discovery of the plant cell.	Future conversation	Modifier
6	Teacher	What do you think **as a scientist, you will need to tell** people?	Future conversation	Projection Modifier
7	Chloe	How it looks like		
8	Teacher	Okay. How the plant cell looks like? And how do you do that?		
9	Mark	Draw?		
10	Teacher	Okay. What else? Yes Andy?		
11	Andy	**You need to tell** people how you discovered it	Future conversation	Modifier
12	Teacher	Ah very good. **You need to talk** about your discovery, like for example where did you discover it? **What other things you need to say?**	Future conversation	Modifier
13		How about **how do you tell it is a plant cell?** What do you need to use?		Sensory
14	Class	Microscope!		
15	Class	(Continue discussing while teacher notes down a list of things to report)		
16	Teacher	So **what you need to do now is... you are**	Ongoing activity	Projection
17		**going to write a simple report like a scientist**	Future conversation	
18		In your report, you **must tell people** about **all these things we just discussed**	Past conversation	Modifier

In this introductory activity, the teacher was using a range of organizational and evaluative metadiscourse to frame the context for the subsequent report writing activity. Notably, projection was used repeatedly in lines 2, 6, and 17 for the students to adopt the voices of a scientist as they "tell people" of their discovery in their report. In addition, modifiers such as "must tell" (lines 5, 18) and "need to tell/talk" (lines 11, 12) were used to signal an almost obligatory way of reporting in their subsequent activity. This informs the students that, as scientists, they have the responsibility to report certain "things" to the public. As the teacher put it in line 12, "What other things you need to say?" Thus, by adopting the identity of scientists and projecting their voices in the report writing, this facilitated the students to understand the rationale of the report from scientists' point of view.

In addition, epistemology markers provide a good opportunity to talk about the epistemic source of scientific knowledge. Teachers and students should develop a habitual way of using epistemology markers when they talk about science, as they signal the evidential basis of their knowledge. This can be a simple addition of a phrase, "according to scientists" to factual statements like "there are eight planets in our solar system." Earlier we saw examples where epistemology markers were used to signal various sources of our knowledge, including sensory experience, logic, postulate, and scientists' work. Admittedly, some of these epistemology markers were implicit remarks and could easily escape students' attention. Thus, there is a need to highlight epistemology markers more often and use them as opportunities to discuss further the nature of science inquiry. For example, in line 13 in the above excerpt, the teacher raised an excellent question concerning what was needed to make the claim of a plant cell and further hinted on the use of equipment to support scientists' observation. The teacher could use this opportunity to emphasize the importance of sensory phenomena and the use of scientific tools and inscription devices (Latour & Woolgar, 1979) as an integral part of scientists' work.

Another way to highlight the use of epistemology markers is to rephrase them using specific technical language, so that instead of embedding them as part of our speech, they become the object of the discussion. For instance, in Excerpt 5.6 when the teacher was performing the water pressure demonstration, instead of saying, "*how do we see* which point has the highest pressure?" to imply sensory experience as the epistemic basis, the utterance could be rephrased into, "what is the *evidence* based on our *observation*?" Also, in Excerpt 5.7, instead of saying "*obviously*, there is *one more chain of answer* here" to imply logic as the epistemic basis, this can be supplemented with another statement like "this will be the *reasoning* in the *explanation*." However, to implement these strategies students need to know the meaning of words like evidence, observation, reasoning, explanation, and how they are connected to scientific knowledge. This will involve the notion of a metalanguage, which I will elaborate in the next chapter.

Fostering Metacognitive Modeling and Development

Lastly, metadiscourse can potentially be used to foster metacognition (cognition about cognition). There are two components of metacognition. The first component is metacognitive knowledge or awareness, which is the self-knowledge of one's thoughts or thought processes in relation to a task. The second component is metacognitive skills, sometimes called self-regulation, which deal with the monitoring and control of one's cognition (Flavell, 1979). New research has found that metacognition can be identified from a person's verbalization that demonstrates explicit expression of one's knowledge or self-regulation (Whitebread et al., 2009). However, the reverse is likely to be true as well; a frequent explicit expression of one's knowledge and self-regulation with respect to a task can lead to the development (or internalization) of metacognition. This postulate is based on Vygotsky's (1986) view of the inseparable connection between language and cognition.

An example of how metadiscourse can foster metacognition is seen earlier in Excerpt 5.7. The teacher's long commentary from line 1 to 8 before he began the explanation reflected his metacognitive knowledge of what was required for completing the task (an O-level examination question) and the procedures needed. This commentary was provided as a form of metacognitive modeling where he made his thinking visible by talking aloud what the students should do before starting the explanation. Such "think-aloud" procedure is commonly used to model metacognitive thinking by many teachers (Fisher, 2002). As we can see in Excerpt 5.7, the think-aloud was facilitated by the teacher's use of metadiscourse. For instance, the activity and text connectives from line 1 to 6 increased metacognitive awareness of the task's requirement. From line 6 to 8, the various evaluative metadiscourse also highlighted the epistemological nature of the answer (as a logic), its importance, and the use of keywords. These are procedures needed in planning an explanation and is a form of metacognitive skill.

Research on the role of metadiscourse in metacognitive development is currently at a very early stage. The integral connection between metadiscourse and metacognition is mainly theoretical and speculative at this point in time (Tang, 2020; Yore, 2018). We need more empirical research to test this connection, and more importantly derive classroom exemplars that demonstrate how teachers can use metadiscourse to foster metacognition. This is an exciting area of research that warrants further investigation.

Summary

Metadiscourse is a neglected area in science education research (Tang, 2017). This is mainly because many researchers and educators are not aware of its existence and usefulness. As a "second-level" discursive resource (to talk about the talk), it tends to escape our attention as we use talk (and text in general) to build the content of our communication. However, metadiscourse is now an emerging area of research in literacy and language education research (Hyland, 2015), with huge potential application in science classroom discourse, as demonstrated in this chapter.

In science classrooms, metadiscourse is particularly useful in crafting a narrative that makes the content of science organized and accessible to an audience as well as conveys the attitudes and stances toward the content. As such, there are two narrative patterns in a science classroom. The first narrative is characterized by the pattern of opening, closing, framing, sequencing, announcing, and summarizing frequently seen in the transition of classroom activities. The second narrative is characterized by a preferred evaluative stance toward the content that is aligned with a particular ideology (e.g., science as empirical inquiry, science as examinable knowledge).

This chapter also introduces a range of metadiscourse strategies that teachers use to provide an ongoing commentary on the unfolding narrative alongside the development of scientific content. The various types of metadiscourse include text, knowledge, and activity connectives for organizational metadiscourse and attitude, epistemology, and interpretative markers for evaluative metadiscourse. Many teachers and students use these metadiscourse strategies unconsciously and indiscriminately in their communication, and therefore miss the opportunity to harness metadiscourse to discern, control, or change the narrative in the classroom. As such, there is a need to help science teachers use metadiscourse more consciously and judiciously to make science content more accessible, organized, relevant, and engaging to learners. In particular, I highlight three discourse strategies that take a more deliberate application of metadiscourse to shape the way students interpret the content of science cohesively and interpersonally. These strategies are (a) narrative framing, (b) heteroglossic projection, and (c) epistemological marker.

Note

1 Here, it is important to distinguish between internal and external conjunctions (Halliday & Hasan, 1976) in order to tell the difference between the use of conjunctions for building ideational content (as a logical relation in Chapter 4) or as an organizational metadiscourse (as a sequencer text connective in Chapter 5). External conjunctions reflect the sequential unfolding of external events in the material world while internal conjunctions reflect a person's rhetorical move to organize the information that is presented. See Tang (2017) for further elaboration.

6

USING DISCOURSE TO

Enact Scientific Practices

We have thus far examined a number of discourse strategies from Chapters 3 to 5 that explicitly support students in learning the content knowledge of science by drawing on the discursive resources of verbal language. These discourse strategies are valuable for students to overcome the abstract and dense nature of scientific language in order to build their understanding of the subject matter. Toward the end of Chapter 5, we also discussed the importance of science inquiry and how evaluative metadiscourse, particularly epistemology markers, can provide useful resources for talking about the nature of inquiry in science classrooms. In this chapter, we will take these ideas around scientific practice further by introducing the notions of genre and metalanguage.

Science education is not just about learning the accepted knowledge of science. It is also about learning how science is carried out as a unique form of human endeavor to understand and interact with the world we live in. Specifically, students need to appreciate how our knowledge of the natural world came about through a set of social practices that are historically developed by the scientific community. This will enable them to become more critical consumers and producers of scientific information. This focus on "practices" is explicitly stated in NGSS in the U.S. as a key dimension in the new curriculum framework besides the conventional emphasis on core ideas and cross-cutting concepts. Clearly, there is an increasing move away from an exclusive focus on the content of science toward a broader understanding of the epistemic origin and underpinning of scientific knowledge.

In the first half of this chapter, the connection between classroom discourse and scientific practices will be elaborated through the concepts of genre and metalanguage. As we discussed in Chapter 2, genre provides the crucial link between language and social practices. At the same time, metalanguage is an explicit form of language to describe the language used in scientific genres. In the second half of this chapter, with the understanding of genre and metalanguage, I introduce specific discourse strategies that enable students to learn two important scientific practices – explanation construction and argumentation.

Genre Patterns of Scientific Practice

The Focus on Scientific Practices

The increasing emphasis on learning scientific practices is most visible in recent curriculum development in the U.S. and other developed countries. A notable shift came from NGSS which defines scientific practices as the "practices scientists employ as they investigate and build models and theories about the world" (NRC, 2012, p. 30). NGSS also explicitly outlines a list of practices that mirror those of professional scientists, namely: (1) asking questions, (2) developing and using models, (3) planning and carrying out investigations, (4) analyzing and interpreting data, (5) using mathematics and computational thinking, (5) constructing explanations, (7) engaging in argument from evidence, and (8) obtaining, evaluating, and communicating information.

The role of literacy has also been recognized as crucial to the development of scientific practices around the world (Tang & Danielsson, 2018). A recent report by the National Research Council (NRC, 2014), titled "Literacy for Science," explored the intersection between NGSS and the Common Core State Standards (CCSS) and made a number of connections between both sets of standards. With the premise that "engagement in the science practices is language-intensive and requires students to participate in classroom science discourse" (NRC, 2014, p. 3), a list of reading, writing, speaking, and listening standards from CCSS was mapped to the practices from NGSS. Table 6.1 reproduces some of the CCSS standards for Grades 9–10 that are related to four of the scientific practices identified as language-intensive:

TABLE 6.1 Examples of CCSS for literacy in science that support NGSS practices (from National Research Council, 2014)

NGSS Practice	*CCSS for Literacy in Science*	*Detailed description from CCSS for Grades 9–10*
Practice 3: Planning and Carrying out Investigations	Following Complex Processes and Procedures	Follow precisely a complex multistep procedure when carrying out experiments, taking measurements, or performing technical tasks, attending to special cases or exceptions defined in the text

(Continued)

TABLE 6.1 (Cont).

NGSS Practice	CCSS for Literacy in Science	Detailed description from CCSS for Grades 9–10
Practice 6: Constructing Explanations	Explaining Concepts and Processes	Write explanatory texts, including the technical processes . . . Develop the topic with well-chosen, relevant, and sufficient facts, extended definitions, concrete details, quotations, or other information and examples appropriate to the audience's knowledge of the topic
Practice 7: Engaging in Argument from Evidence	Making Arguments	Develop claim(s) and counterclaims fairly, supplying data and evidence for each while pointing out the strengths and limitations of both claim(s) and counterclaims in a discipline-appropriate form and in a manner that anticipates the audience's knowledge level and concerns
Practice 8: Obtaining, Evaluating, and Communicating Information	Translating Information from One Form to Another	Translate quantitative or technical information expressed in words in a text into visual form (e.g., a table or chart) and translate information expressed visually or mathematically (e.g., in an equation) into words

From Table 6.1, it can be seen that each of the NGSS scientific practices is built upon a particular way of using language to frame the activity of the practice and consequently produce the texts associated with that practice. For instance, in planning and carrying out investigations (Practice 3), there is a certain patterned way of reading, writing, and executing the steps (i.e., procedures) needed for the experiment. Similarly, there is a pattern in constructing scientific explanations (Practice 6), which are defined by technical processes, definitions, and relevant facts in explanatory texts. I will later elaborate these patterns through the lens of a genre.

As a prelude, there are four major genres in scientific discourse – information report, experimental report, explanation, and argument (Halliday, 1993a). Practices 3, 6, and 7 as defined by NGSS are associated with the genres of experimental report, explanation, and argument respectively. However, Practice 8 is not associated with any particular genre. Instead, this practice is embedded within all the major genres of science. For instance, the literacy act of "translating information from one (multimodal) form to another" is needed in the genres of report, explanation, or argument. Yet, the characteristic of this multimodal translation is dependent on and varies across the different genres. These distinctions in Practices 3, 6, and 7 will be further elaborated in this chapter, while the distinction in Practice 8 will be elaborated in the next two chapters on multimodal discourse.

The Role of Genre: Connection between Text and Social Practice

In Chapter 2, I have described a genre based on the theory of SFL as a culturally evolved way of doing things with language. Genres refer to the predictable patterns and conventions that are associated with and partly realize the repetitive activities that people do in their social life (Fairclough, 1992). They are also the intermediary between text and social practice. There are numerous genres in any culture. Every child learns to recognize and mimic the common and simple ones such as telephone call or shopping transaction by noticing recurring patterns of using language in our daily interactions with others. As these patterns are quite consistent in a given culture, this is how we can learn to predict how an activity is going to unfold, and consequently know how to interact in it. Similarly, when these patterns are disrupted, we will notice something is odd with the situation and find it difficult to respond appropriately.

Such distinctive patterns are characterized by how language is used to mediate different stages of a social activity in order to fulfill its completion. Over time, the patterns evolve and develop into a set of conventions that is not only recognizable within a community, but also becomes a kind of framing that guides our participation within familiar genres. A genre can generally be identified from three characteristics: social purpose, stages, and functions (Martin & Rose, 2007). First, a genre is always organized around a common purpose of achieving a particular task with someone (e.g., receiving information, buying things). Second, achieving this purpose often involves several steps or stages that have stable and predictable patterns of organizations. Third, each of these stages serves a particular meaning-making function, which is achieved by distinctive language and interaction variations (or a pattern of register; see Chapter 2). For instance, in the product enquiry stage of non-online shopping, the typical pattern is the consumer asking questions and the salesperson giving information centering on product specification, pricing, and durability. In the transaction stage, however, the pattern shifts to the salesperson requesting information or offering products centering on finances and payment.

The pattern arising from the three above-mentioned characteristics is how we can identify genre. This pattern is easily recognizable in written texts, which is more systematic and organized compared to oral interaction. For this reason, it is useful to first examine different genres within a written mode, and later extend the idea to incorporate other modes in classroom discourse. In scientific written texts, functional linguists have identified four major genres (Halliday, 1993a). These genres, with their corresponding purpose and functional stages, are summarized in Table 6.2 (adapted from Martin & Rose, 2007).

Besides written texts, these genres are also found in oral interaction. In science classrooms, the four scientific genres play an important role in shaping the teaching and learning of science. Many students are not familiar with the genres of science and how they affect what is considered as acceptable

TABLE 6.2 The purpose, stages, and functions of scientific written genres

Genre	Purpose	Stages and Functions
Experimental report	To present the procedures and results of an experiment	Aim – objective, question or hypothesis Methods – procedures taken and apparatus used Results – conclusion from data analysis
Information report	To organize information about things or events in the world	Classification – different types of entities Compositional – parts of an entity Description – attributes and properties of entities
Argument	To state a claim or position and present supporting evidence in favor of the claim or position	Thesis – position or stance Argument – justification of position Discussion – consideration between two positions
Explanation	To account for the underlying causes or processes of a phenomenon	Phenomenon – what is being explained Sequence – series of causes and consequences

knowledge in science. The following excerpt from a Grade 5 classroom illustrates this point where the students could not follow the genre of a scientific explanation in this interaction:

EXCERPT 6.1

#	Speaker	Utterance
1	Teacher	I already explained to you. **How** is shadow formed?
2	Qian	When light is blocked
3	Teacher	Ben, **how** is shadow formed?
4	Ben	When the something blocks . . .
5	Teacher	What is the something?
6	Ben	Object
7	Teacher	No. Okay, when I ask you **how** is shadow formed, you cannot say when something is block by the object. **What else?**
8	Ahmad	Blocked by an opaque object
9	Teacher	**So how?** My question is *HOW* is the shadow formed? (*writes on board*)
10	Teacher	So when light is blocked by an object? **So what?**

At the beginning, the teacher wanted her students to reiterate the explanation of "how is shadow formed?" which she had given earlier before Excerpt 6.1 occurred. In this interaction, the students were able to follow and participate in the IRE/IRF interaction pattern (see Chapter 3 for IRE and IRF patterns).

They were also able to give relevant information that was not technically wrong (e.g., blocked by an opaque object). However, it was obvious that the teacher was not satisfied with all the answers given by Qian, Ben, and Ahmad. By examining the way the teacher repeatedly stressed the "how" aspect of the explanation after every response given (lines 1, 3, 7, and 9), we can see that she knew the explanation given so far was not sufficient.

So why was the explanation given by the students not sufficient? Going back to the definitions of explanation by NGSS and CCSS (Table 6.1) and the genre of explanation (Table 6.2), what was missing here is the "concepts, causes, and processes" behind the explanation. In line 10, the teacher tried to give some hints through an *extend* follow-up move – "so what?" As discussed in Chapter 3, this follow-up question aimed to prompt the students to build the processes and cause-effect reasoning of the explanation. From her questioning, we could thus tell that the teacher knew about the characteristic of a scientific explanation. But she only knew it *implicitly*, and she had no other way of communicating that knowledge to her students other than using questioning and repeating "how" five different times, in lines 1, 3, 5, and 9.

At the same time, the students had never been *explicitly* taught what an explanation is, although they had heard this term numerous times from teachers, textbooks, and assessment materials. Thus, it is not surprising they were unable to follow the explanation genre we saw in this typical excerpt. This is a very common problem encountered in many science classrooms. What is needed is a discourse strategy that will *make explicit* the genres of science. Such discourse strategy will involve the use of a metalanguage.

The Role of Metalanguage: Making Explicit the Genres of Science

To explicitly talk about scientific genres will involve teachers and students learning a language to talk about the function and structure of language found in those genres. In other words, they need a metalanguage. The New London Group (1996, p. 77) defines metalanguage as a "language for talking about language, images, texts, and meaning-making interactions." It is a kind of "second-level language" that is not normally used by the people involved in their own discourse practices, but serves as a crucial communicative tool to describe and reflect on those practices (Shanahan, 2012). For example, words like *subject*, *verb*, *clause*, and *conjunctions* are seldom used in our everyday usage of English, yet they are important words used by linguists and language teachers, or when we need to reflect on our own or someone's use of English. Thus, metalanguage is used as a self-referential tool of one's own language and this is where the prefix "meta" comes in.

In science, examples of metalanguage include words like *claims*, *evidence*, *law*, *model*, *theory*, *hypothesis*, and *prediction* (Norris et al., 2008). These words are not commonly used by scientists in their own research and communication.

However, they are necessary for historians, philosophers, and sociologists of science to analyze and describe the nature of scientific work. Some scientists do use these words in their work, but usually for the purpose of communicating with the public and not within the discourse of science, like presenting at scientific conferences or publishing in journals. In science education, a number of researchers have argued that the metalanguage of science plays an important role in learning the epistemology of science and should be used more often during science instruction (e.g., Norris et al., 2008; Shanahan, 2012).

The reason why most scientists do not use metalanguage in their discourse (unless they are writing to a general audience) is because they have already mastered and internalized the language of science. For example, in a scientific article, the meanings of claim and evidence are already implied through the conventions found in the genre, and it is therefore unnecessary for scientists to use these terms to explicate the meaning of their statements. In other words, the meaning of these words and their associated practices are implicit in scientific genres and thus obvious among scientists. One could also argue that scientists have the metacognitive awareness to recognize and reflect on these implicit meanings as they read scientific journals (Tang, 2020). However, the same cannot be said for novices who are learning not only a new language of science, but also new forms of practices (e.g., scientific argument and explanation) that are sometimes at odd with the everyday practices they are used to (Moje, Collazo, Carrillo, & Marx, 2001). Thus, it is important for children to learn scientific metalanguage as an explicit form of language to describe and analyze the way language is used in scientific practices. For this reason, the New London Group (1996) argues that metalanguage is needed as the "languages of reflective generalization that describe the form, content, and function of the discourses of practice" (p. 34).

Metalanguage Discourse Strategies

Among the various genres, the metalanguage that is most visible to science teachers and students is the one associated with experiments, through words like *aim, method, procedure, independent/dependent variable, observation,* and *result.* In many countries' high school science curricula, this metalanguage is often taught and used during science practical or laboratory periods (which is usually given a dedicated time in the curriculum besides the main "theory" lessons). However, the same could not be said for the other scientific genres.

In this chapter, I will focus on the metalanguage of two particular genres – scientific explanation and argument. There are two reasons for this focus. First, these genres are more subtle and challenging to teachers and students as compared to experimental report and information report. Second, these two genres are crucial for students to understand the epistemic nature of science. As such, I will expand on the discourse strategies that incorporate the metalanguage of scientific explanation and argumentation.

The PRO Metalanguage

The metalanguage for scientific explanation addresses the problem we saw from Excerpt 6.1 where the teacher could not get her students to sufficiently explain "how is shadow formed?" despite getting a number of correct responses from them. Although the purpose of that activity was to form an explanation, there was no clear criteria or guideline to the participants as to what exactly is a scientific explanation.

The word *explanation* has multiple meanings in the context of science teaching, and for our discussion, it is important to distinguish between *scientific explanation* and *pedagogical explanation* (Horwood, 1988). Pedagogical explanation arises from the loose usage of the word "explain" whenever an explication is required to unpack a scientific idea in order to provide some clarification or promote students' understanding (Braaten & Windschitl, 2011). Very often, it is an elaboration of a scientific term, concept, or theory with the use of techniques like repetition, paraphrase, analogy, or recalling prior knowledge in the elaboration. On the other hand, a scientific explanation has a natural phenomenon that needs to be explained or accounted for through some logical and causal processes. This definition of explanation is stated in NGSS as "the explicit applications of theory to a specific situation or phenomenon" (National Research Council, 2012, p. 52). As an example, "explain what is a light ray" is a pedagogical explanation that expands upon a scientific concept called light ray (which is technically a model in this case), but it is not a scientific explanation because there is no natural phenomenon involved. By contrast, "explain using light ray why or how a shadow is formed" is a scientific explanation because one can observe a shadow.

Philosophers of science define scientific explanation as a theoretical or mechanistic account of why or how natural phenomena occur the way they do (Achinstein, 1983). A scientific explanation typically involves a number of characteristics. First, it involves a phenomenon that either is observable or has already occurred according to our current knowledge. Second, the account typically involves a causal mechanism or temporal sequence that leads to the phenomenon observed. In Chapter 4, we saw how this causal mechanism consists of a chain of logical relations (joined by conjunctions, e.g., because, thus, next) that builds up the "causation" of the explanation (Unsworth, 2001a). Third, scientific explanations are often constructed logically based on regular patterns that derived from and are validated through repeated observations (e.g., laws) or a model that connects seemingly disconnected phenomena with a unifying framework or theory (Friedman, 1974; Hempel & Oppenheim, 1948).

Based on the three above-mentioned characteristics, I developed a metalanguage based on a research study to describe the genre of scientific explanation, using three technical terms called *premise, reasoning,* and *outcome*.[1] Premise (P) is an accepted or assumed knowledge that is used as the theoretical basis or "first cause" of an explanation, and it includes scientific definition, law, principle, model, theory, postulate, and fact.[2] Reasoning (R) is the logical sequences that follow from the premise, and outcome (O) is the phenomenon to be explained. (See Tang (2016a) for more background on how this metalanguage was developed from the research study.)

Relating back to the explanation on "how is shadow formed?", Table 6.3 is an example of how the explanation will look based on the PRO metalanguage:

TABLE 6.3 A scientific explanation of "how is shadow formed?" based on PRO

Metalanguage	Explanation Sequence	Remarks
Premise	• Light travels in a straight path from a light source	This is one of the basic postulates from the ray model of light. There are alternative theories, such as the wave or particle theory of light, but these are more complicated theories that are not necessary for this explanation.
Reasoning	• *When* an opaque object blocks the light path between the source and a screen, • the light cannot continue traveling to the screen	This reasoning can and should be strengthened with the use of a diagram to show (in a spatial sense) how the light path is blocked from the source to the screen. The integrated use of visual representation within the genre of explanation will be discussed in Chapter 8.
Outcome	• *Therefore*, a shadow (which is the absence of light) will be formed	The chain of reasoning that links the premise to the outcome is formed by conjunctions as italicized (i.e., when, therefore).

As we recall in Excerpt 6.1, the teacher was not satisfied with her students' responses as they only mentioned about the opaque object blocking the light. They did not state the premise nor extend the reasoning of light path traveling from source to screen. These two aspects are normally associated with the "concepts" and "processes" of scientific explanation respectively (see definition of explanation from CCSS). Using the PRO metalanguage, this will help the teacher and her students to discuss more explicitly what was missing in their incomplete explanation.

Explanation Construction Using PRO

Teaching PRO as a Writing Heuristic

In order to incorporate PRO into classroom discourse, teachers and students need to first learn the PRO metalanguage explicitly. Ideally, PRO should be taught to students as a heuristic at the beginning of an academic year and subsequently reinforced throughout the year for different topics. This was the approach taken in several studies that pioneered the use of PRO (e.g., Lee & Goh, 2017; Putra & Tang, 2016; Rappa & Tang, 2018; Tang, 2015, 2016a). The introduction of PRO should be done in the context of an instructional task anchored by a puzzling phenomenon from an interesting demonstration or experiment. Instructional models, such as the 5E inquiry (Bybee et al., 2006) or the Predict-Observe-Explain (POE; White & Gunstone, 1992), are suited in setting the instructional context for the introduction of the PRO. In particular, the E for "Explain" in both the 5E and POE is where PRO comes in handy as many students tend to struggle in knowing how to write an explanation.

In the following example, a teacher introduced the PRO metalanguage to the students for the first time in a lesson on ionic bonding. Following the 5E inquiry approach, a demonstration was used to show that some compounds (e.g., potassium chloride) can conduct electricity in the molten state but not in the solid state. After the class had discussed their initial ideas, the teacher used PRO with the goal of helping her students to write the explanation based on their earlier discussion:

EXCERPT 6.2

#	Speaker	Utterance
1	Teacher	This is the PRO framework. What does PRO stand for? Professional, eh? Alright, so if you want to be very professional in writing, in crafting your answer, we need to look at number one, the **Principle**.
2	Teacher	**Principle** simply means what you know about this particular bond. For example, the structure and the bonding. You can state for me what kind of bonding is it, how is it being held together. The forces of attraction. Alright?
3	Teacher	R is the **Reason**, alright? So for anything that happens, for anything that you observe, there must be a reason. Why is it when solid, it didn't light up, but when molten, it lights up. Alright?
4	Teacher	And **Observation** is what you see, alright?

In this example, because this was the first time students heard about the PRO metalanguage, the teacher was explicit in line 1 in describing this new strategy and explaining its purpose as a writing heuristic (e.g., "PRO framework," "crafting your answer"). Subsequently, she used specific examples to describe the strategy rather than giving a general definition of PRO. This helped to make it easier for the students to understand. Thus, the use of "principle" (line 2) instead of "premise" and "observation" (line 4) instead of "outcome" was applicable and relevant in this case. However, it is important to note that the terms "premise" and "outcome" are more general and therefore more applicable in other scientific explanations when there is no clear principle or when the outcome is not directly observable.

Developing Non-linear Reasoning

The PRO metalanguage is not only useful as a heuristic to support student writing, but it can also be used to facilitate non-linear reasoning. Most science teachers tend to develop scientific explanation linearly from beginning to end. For instance, we saw in Chapters 3 and 4 how teachers conducted whole-class discussions using various interaction and thematic discourse strategies (e.g., IRF questioning, unpacking abstraction). In almost all the excerpts, we find that the logic of the unfolding explanation tended to occur in a linear cause-effect manner.

The PRO metalanguage provides a useful method to change the unfolding of the explanation into a non-linear sequence. In this example, the students were performing a "dancing" raisins demonstration by putting some raisins into a beaker of soda water. After they observed the raisin sunk and floated repeatedly for a number of times, the teacher got the students to discuss the explanation for this phenomenon. Instead of developing the explanation from beginning to end, he broke down the reasoning structure of the explanation by writing a series of P, R, and O on the board (see Figure 6.1). He then integrated the PRO metalanguage into his whole-class discussion as shown in Excerpt 6.3:

EXCERPT 6.3

#	Speaker	Utterance	P/R/O
1	Teacher	Okay, let's explain what we **observe** . . . usually by **principle** . . . you can state the formula or definition of density. . . . Density is mass per unit volume.	P
2	Teacher	What did you first **observe**? Let's write down the **observation**. The raisin, the moment you put it in, actually most of them actually?	O
3	Class	Sink	O
4	Teacher	Sink. So **outcome** is the raisin sink. So why do you think the raisin will sink?	R
5	Wei Lee	Due to fact that they are more dense than (inaudible)	R
6	Teacher	Correct. Due to the density of the raisin is higher . . . And so your raisin sink.	R
7	Teacher	After that, what do you **observe**? The raisin actually?	O
8	Rami	Float	O
9	Teacher	Why did the raisin float?	R

(*Continued*)

EXCERPT 6.3 (Cont).

#	Speaker	Utterance	P/R/O
10	Nichole	Cos carbon dioxide …	R
11	Teacher	Ah then, shouldn't the carbon dioxide just floats on its own? … Now what actually happens to the carbon dioxide gas bubbles?	
12	Jamie	Attach itself to the raisins	R
13	Teacher	Yes, it attached itself to the raisins. The gas bubbles attached themselves to the raisins.	R
		And when it is attached to the raisin, what happens to it? What increases?	R
14	Nichole	Volume	R
15	Teacher	The volume increases … So what happen when the volume increases?	R
16	Kris	Average density decrease	R
17	Teacher	The average density decreases. The average density decreases. So, so what's the **outcome**? The raisin with the air bubble?	O
17	Karen	Floats	O

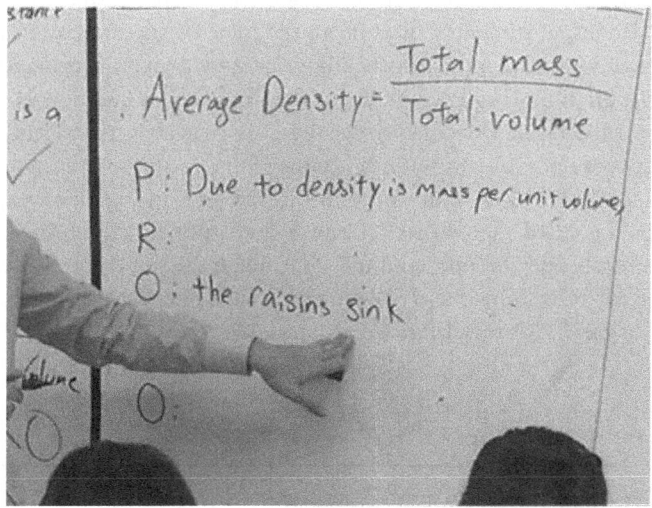

FIGURE 6.1 PRO used as navigational markers in the dancing raisins explanation

As we see from the excerpt, the teacher first established the premise for the explanation, and through his questioning technique (predominantly IRF; see Chapter 3), got the students to give, in stages, each outcome before giving the reasoning behind it. Thus, instead of a linear cause-effect sequence, he used a P-O-R-O-R-O sequence in developing the explanation. This was made possible through the shared metalanguage where the P, R, and O were used as navigational markers to facilitate the interaction between the teacher and his students and regulate their key ideas back and forth between different parts of the dancing raisins explanation. The writing of P, R, and O on the board (Figure 6.1) also provided visual markers to guide the students in the reasoning process.

The non-linear development of a scientific explanation is important as it unpacks the epistemic structure and pattern of a scientific explanation. This helps students to see how an explanation works; not just in terms of a causal or temporal sequence but also in terms of the connections among the various epistemic components (e.g., premise, reasoning, outcome). More importantly, the use of the PRO metalanguage allowed teachers and students to explicitly talk about these connections. As a result, it also functioned as an epistemic tool that supported the students to make their thinking visible and aligned with scientific practices (Tang, 2020).

Evaluating Explanation

A metalanguage also provides the tools for students to evaluate their construction of scientific explanation by making reference to their own language as they reflect and critique on the ongoing development of an explanation. Equipped with a metalanguage, students can evaluate the logic of an explanation according to its genre, and consequently articulate how or why a given explanation is valid or invalid. In the following excerpt, a student first raised a question of "why hot air rises?" and the teacher decided to let the class answer it. A student, Nguyen, initially proposed an explanation with a rather long and sound reasoning. However, a few students followed up to point out the gaps in Nguyen's explanation. This excerpt was briefly shown in Chapter 3 (Excerpt 3.13) to illustrate students' constructive challenges to one another. The full transcript is now presented in this excerpt to show how their constructive challenges were made possible by their usage of the PRO metalanguage:

EXCERPT 6.4

#	Speaker	Utterance
1	Nguyen	I was thinking of the comparison between the cool air and the density of hot air. So when you compare, hot air will rise because lesser density . . . Normally you will say the denser one will sink. So since hot air rises, means the density of the hot air is lower than the density of the normal air. Since normal air is having a . . . is denser. So it will sink. So when it sinks, the hot air rises due to the density is lower. It's like you compare with a liquid and cork . . . .

(Continued)

EXCERPT 6.4 (Cont).

#	Speaker	Utterance
2	Teacher	Okay, that's his **reasoning** right. So his main point is that, he straightaway thinks of the outcome, hot air rises. He thinks there must be something that cause it to rise or sink, that is due to density difference . . . That's why he went on to say density decreases like what Jen said . . .
3	Rhoda	What's the **principle?**[1]
4	Teacher	Yah, it's already stated here (points to the written **P**)
5	Jihoon	But it's not related to
6	Farid	It's not link to any **principle**
7	Teacher	Correct he has another **principle**. Do you know what's the **principle** he was talking about just now?
8	Feng	Density difference
9	Teacher	Yes, density difference. That was his **principle**. Now Jay you were saying something?
10	Jay	If you want to **reason**, you have to ask why does the density decrease, then go backwards.

1 The class used the term "principle" instead of the more general term "premise" during the study.

In this excerpt, Nguyen gave a rather impressive explanation in line 1 that was both conceptually and logically accurate. This explanation involved a logical reasoning predicated by a long chain of causal sequences (i.e., "hot" causes "lower density" causes "rises," and "cooler" causes "denser" causes "sink"). We then see three students, Rhoda, Jihoon, and Farid, objecting to this explanation in lines 3, 5, and 6. Their objections, or constructive challenges, were not directed at the conceptual accuracy of Nguyen's explanation. Instead, they took issue with the epistemic nature of the explanation in terms of lacking a reasonable principle (or premise in general). Note that it was the use of the metalanguage that made it possible for them to explicitly articulate the reason for their objections. Subsequently, this facilitated the class to discuss there were different premises, and the premise behind Nguyen's explanation – density difference – was not sufficient to answer the question (line 7 to 9). As Jay then suggested in line 10, they needed to reason backward to arrive at a more fundamental premise (the kinetic model of matter), which the class eventually did.

If we compare this excerpt with the earlier Excerpt 6.1 (on how shadow is formed), we can see a notable difference. In Excerpt 6.1, when the students could not state the underlying premise, the teacher could not explicitly tell them what was wrong, other than saying "why" or "how." Such feedback lacks the precision of identifying the crucial component in the explanation genre. When the teachers and students learned a shared metalanguage, the class now possessed the meaning-making resource for them to reflectively analyze and discuss the genre of the explanation they were developing. This is a main reason why metalanguage plays a very important role in classroom discourse, particularly in learning the genres of the discipline.

The CDW Metalanguage

The second metalanguage we will examine now focuses on scientific argument, which is closely related to scientific explanation but has a very different epistemic function (Osborne & Patterson, 2011). While an explanation seeks to apply theory to a specific situation or phenomenon (National Research Council, 2012), an argument seeks to persuade others by justifying a claim or position in the light of supporting or contradictory evidence. According to the philosopher Toulmin (1958), an argument minimally must consist of a *claim (C)*, which is a tentative proposition, *data (D)* as the evidence to support the claim, and *warrant (W)*, which connects the data to the claim. An argument will sometimes include a *rebuttal (R)* that seeks to question or challenge other claims. Similar to an explanation, the mechanism that connects the data to the claim consists of a chain of logical relations (joined by conjunctions, e.g., because, thus, next) that builds up the "rhetoric" of the argument.

In terms of structure, an explanation and argument are quite similar; the structure of premise-reasoning-outcome is comparable to data-warrant-claim. However, they differ significantly in terms of the tentativeness of the account to be explained or argued. In an explanation, the phenomenon to be explained (i.e., outcome) is not in doubt or has already occurred. For example, questions such as "how does an airplane fly?" or "why did dinosaurs become extinct?" are explanation questions because the phenomena (i.e., an airplane can fly, dinosaurs are extinct) are not in dispute (Osborne & Patterson, 2011). In other words, the outcome is definite and not debatable. An argument by contrast has a degree of uncertainty over the claim to be argued, without which there would be no argument. As such, there must be at least two competing claims in order for argumentation – the dialogic *process* of making argument – to take place (Duschl & Osborne, 2002).

Another reason why explanation and argument are often conflated is because the persuasiveness of an explanation requires argumentation, and conversely, the process of argumentation involves multiple explanations. Take as an example, "how does a gigantic asteroid collision kill the dinosaurs?" To address this question requires an explanation based on the premise of an asteroid collision, the causal consequences of releasing dust from the impact and creating an atmospheric cloud, and the undisputable outcome that dinosaurs are extinct. The logic of this explanation will be judged according to the extent the account is coherent, both internally and externally (Thagard, 2008). Internal coherence is based on the logical causality of the explanation from the premise; for example, collision led to release of dust, which led to an atmospheric dust cloud, and so on. External coherence is determined by accepted scientific knowledge that supports the causal reasoning, such as the composition of a dust cloud and why it will block sunlight. Based on internal and external coherences alone, there is scientific consensus that a large asteroid collision *will* end the dinosaurs.

But the real question is whether there was a large asteroid collision at the end of the Cretaceous Period. This question cannot be answered within the explanation itself. In other words, the explanation cannot be judged by whether the asteroid impact did actually happen. To address this question will require an argumentation on the premise of the explanation, which is that there was a large asteroid collision in the first place. The argumentation will require supporting empirical evidence, such as the Chicxulub crater and the high concentration of iridium at the Cretaceous–Paleogene rock boundary. It will also involve alternative explanations that are as likely and have not been ruled out yet, for example volcanic eruption and climate change. Therefore, argumentation involves multiple explanations in contention. At the same time, the persuasiveness of an explanation depends on whether this explanation is better than another alternative explanation in the argumentation process, and not within the internal or external coherence of the explanation itself.

What is the argument for an asteroid collision killing the dinosaurs? Table 6.4 gives an example of an argument for the "asteroid theory" based on Toulmin's argument structure:

TABLE 6.4 A scientific argument for the "asteroid theory"

Metalanguage	Argument Sequence	Remarks
Data	• Large crater at Chicxulub • High iridium level at Cretaceous–Paleogene (K-Pg) rock boundary. Iridium is rare in Earth's crust but abundant in most asteroids	These are two of the most common pieces of empirical evidence for the asteroid theory
Warrant	• Based on the crater diameter, the asteroid is estimated to be 10 km wide • Based on the K-Pg boundary, the event occurred during the Cretaceous period	
Claim	• An "extinction level" asteroid hit the Earth during the period when dinosaurs lived	This theory, though widely popular, has not been universally agreed by geologists and paleontologists as there are currently other competing theories

In the last decade, there has been a number of intervention studies that apply Toulmin's (1958) model of argument to develop a three-part rhetorical framework to help students engage in scientific argumentation (e.g., Erduran, Simon, & Osborne, 2004; McNeill & Krajcik, 2008; Sandoval & Millwood, 2005; Wang, 2014). These are examples of using a metalanguage to describe the language of scientific argument (as a product) and argumentation (as a process). In these studies, the teachers learned the metalanguage and taught it to their students explicitly. There are slight variations in the choice of words for the metalanguage; for instance, McNeill and Krajcik (2008) use the terms claim, evidence, and reasoning while Erduran, Simon, and Osborne (2004) use claim, data, and warrant, following Toulmin's original wording. For this chapter, I will use the acronym CDW (claim-data-warrant) to denote the metalanguage for scientific argumentation.

Argumentation Using CDW

Similar to PRO, CDW provides a useful metalanguage to help students write a coherent scientific argument. However, its usefulness extends beyond just a writing frame to include supporting students in the process of argumentation. In this aspect, CDW is best used as a discourse strategy to promote in-depth discussion and debate among the students. This is because argumentation is fundamentally a dialogic event carried out among two or more individuals (Kim & Roth, 2014). In Chapter 3, we have discussed various ways and challenges in promoting student dialogue in a small group discussion. One of the challenges is that group discussion, even when the students are cooperative, tends to revolve around sharing factual information and acknowledging one another's responses. By contrast, constructive challenges to each other's ideas are rare in group discussion. Even when students do challenge one another, they often lack the discursive resources to articulate their objections and convince their peers the validity of their ideas. The use of CDW metalanguage in the students' discourse is a way to overcome this challenge.

Competing Theories Argumentation

The most important requirement for an argumentation discourse strategy is for students to consider multiple explanations instead of a singular account. One instructional approach is to use what Osborne, Erduran, and Simon (2004) call "competing theories argumentation." Students are given two or more "theories" as claims they will need to agree or disagree with. They are then given the data (in the form of written statements, pictures, videos, or reading materials) that can support or contradict one of the claims, and asked to think of the warrant to support their claim. The CDW metalanguage needs to be explicitly mentioned and used in the instructions for the students. To foster student-student interaction, each student can also be asked to take a stance and be placed in a group with students from an opposing view. Jigsaw cooperative strategy (where students are first placed in an "expert" group to research on a similar topic or position, and subsequently divided into "diverse" groups for debate) are ideally suited for such discursive activities.

For example, an interesting lesson can be designed on "what killed the dinosaurs?" There are four suitable theories that are competing as the most likely explanation – asteroid collision, volcanic eruption, climate change, and epidemic spread. There are a lot of good websites where the students can read about each of these theories. The CDW metalanguage will provide a useful scaffold for the students to identify the data and warrant in the readings, as well as evaluate the gaps in their opponents' data and warrant. Once the students prepare their arguments and rebuttals in their groups, they can then engage in argumentation through small group or whole-class discussion.

In the earlier example on "how is shadow formed?" in Excerpt 6.1, the statement "light travels in a straight path from a light source" is an accepted premise used to explain the formation of shadow. Based on the PRO framework, the premise is taught to students as an accepted or assumed knowledge that they can use as the basis of the explanation. However, that does not mean that they do not question the validity or source of this premise. Complementing the explanation lesson on light, an argumentation activity can be designed for students to debate about the premise. For example, two competing claims can be: (1) light always travels in a straight line and (2) light can bend around an object. Optical objects such as torchlight, pinhole, prism, mirror as well as pictures of natural phenomena such as sunset, blood moon, and various optical tricks can be used as data to support both claims.

Competing Predictions Argumentation

Competing theories argumentation works around contentious theoretical statements or hypotheses that cannot be explained, because they are in fact the premises of explanation. In the primary and secondary school science curriculum, there are not many of such statements that are suitable for the students to practice argumentation. Thus, another common method of designing argumentation activities revolve around observable phenomena rather than theories. However, in order to introduce a degree of uncertainty for the purpose of argumentation, the phenomenon is not shown or demonstrated prior to any argumentation; otherwise, there will be no debate. In other words, the claims will consist of multiple predictions of a phenomenon, and the claims will remain uncertain until a demonstration is performed to resolve the ambiguity. I call this approach a competing predictions argumentation.

The following lesson example was designed around a competing predictions argumentation on chemistry qualitative analysis (QA). A mysterious water was given to students and they were asked to predict if it was contaminated, and if so, what chemicals were found in the water. In groups, students were then asked to conduct a series of tests to obtain the data for their claims. At the end of the lesson, each group presented their arguments to the class and other groups were encouraged to raise questions or objections to their peers. The excerpt below shows an example from a group's oral presentation using the CDW metalanguage (C – claim, D – data, W – warrant, R – rebuttal):

EXCERPT 6.5

#	Speaker	Utterance	C/D/ W/R
1	Sue	Our group's **claim** is that the water is contaminated because there are lead ions in the water	C
2	Thila	When potassium iodide was added, an insoluble yellow p.p.t. (*precipitation*) was formed. This is our **data**.	D
3	Xiao Feng	So based on our test, the potassium iodide reacts with the lead to form lead iodide. And the lead iodide is the p.p.t.	W
4	Nadia (from another group)	But it can be aluminum too right?	R
5	Xiao Feng	No. But aluminum iodide is soluble what. So you won't see the p.p.t. ...	W
6	Thila	Yah, the p.p.t. is the lead	W
7	Teacher	Nadia, what did you get?	D
8	Nadia	We also got yellow p.p.t. But we didn't get lead	D
9	Teacher	So what is your group's claim?	C
10	Nadia	Aluminum. So it's okay to drink	C
11	Teacher	Okay. So do you see why your group has a different **conclusion** even though you have the same **data**? ... How about the rest? Who do you think is correct here?	C D

In this example, the CDW metalanguage was first used to support the students in preparing their arguments and subsequently integrated into their oral presentation and class discussion. As seen from line 1 to 3, the group broke down their results into three portions according to the claim-data-warrant structure, and a different member spoke on each of the portions. A student from a different group then spoke up to offer a rebuttal that the same data (yellow p.p.t.) could be interpreted as aluminum (line 4), which would then lead to a different claim. However, this was met with a counter-rebuttal from the group (lines 5 and 6) who strengthened their warrants by arguing (correctly) that as the possibility of aluminum was eliminated by its solubility, the only option left to conclude was lead iodide. This argument was subsequently resolved by the teacher who pointed out that they could get "a different conclusion even though [they] have the same data" (line 11).

The metalanguage in this example (e.g., data, claim or conclusion) was used to facilitate the conversation, not only in terms of presenting the group's finding according to its argument structure, but more importantly in terms of pointing out how different claims could be inferred from the same data. Just like the use of PRO metalanguage for explanation, CDW was useful in getting the students to evaluate and specify which part of the argument is wrong. In addition, it also helped them to become familiar with the genre of argument and use it to engage in the practice of argumentation, notably in generating and using data to support/contradict a claim through the warrant.

Scientific Explanation and Argumentation in Unison

Scientific explanation and argumentation form a tight relationship in the production, evaluation, legitimization, and communication of scientific knowledge. In order for students to gain a deep understanding of the epistemic nature of science through its social practices, it is important for them to learn both of these genres and their corresponding metalanguage. As mentioned earlier, the persuasiveness of an explanation requires argumentation, and conversely, the process of argumentation involves multiple explanations. There are opportunities for students to learn about this relationship between explanation and argumentation through the examples I have provided earlier.

For example, we saw that "light travels in a straight path" can be both a premise to explain observable phenomena and a claim to be argued in light of alternative claims. Using the metalanguage of premise and claim will allow students to understand that some statements like law, theory, and principle cannot be explained or proved. Instead, scientists derive these statements based on abstraction of observed regularities and use them as a premise to account for or predict a specific phenomenon. In this case, the reasoning that links the premise to the phenomenon is based on logical deduction. Although a premise cannot be explained, it can be questioned and challenged. In science, unlike other disciplines, the authoritative way to challenge a premise is through empirical evidence. In this sense, a premise is therefore also a claim that must meet two conditions. First, it must be supported by data (more is better) through a warrant, and second, it is subjected to be challenged by counterclaims.

The relationship between premise and claim should also be examined further in light of the nature and history of science. As was discussed in Chapter 2, scientists produce claims in journal articles based on the evidence they gathered in a laboratory or fieldwork (Latour & Woolgar, 1979). The strength of the claims they can make depends on a network of "allies" they can assemble, including collecting more data, convincing their peers, receiving funding and citations, winning awards, and generating public buy-in. As the network of allies becomes larger, the claim becomes stronger and more convincing. Eventually, when nobody in the scientific community disputes the claim anymore, then the claim will become accepted as a scientific "fact." In other words, claims can become premises depending on the consensus from the scientific community.

Conversely, many premises that we now accept as facts were historically claims that were debated among scientists. For example, facts that we take for granted like "light travels in a straight line from a light source," "matter is made of particles that are in constant and random motion," and "dinosaurs once roamed the earth and are now extinct" were once competing with other theories at one time in the history of science. Similarly, claims like "an asteroid collision killed the dinosaurs," which are in contention today, may be accepted as a fact in the future when it gathers more allies over other competing theories. The historian of science Thomas Kuhn (1962) gave a pragmatic picture that this will usually happen after a few generations when the reigning scientists who support the opposing theories have retired or passed on.

At the same time, when a claim becomes accepted as a premise, it is always falsifiable (Popper, 1963). This means that it can still be rejected in light of any new phenomenon that cannot be explained by the premise, which Kuhn (1962) calls an anomaly. When such anomaly surfaces, the premise will be questioned and weakened back to a claim to be evaluated against supporting and contradicting evidence. This has happened many times in the history of science. For instance, the premise that "light travels in a straight path" is based on the ray model of light, which is a 17th century invention in the development of geometrical optics. Along with other premises from this model (e.g., law of reflection, Snell's law), they can be used to explain a range of optical phenomena. However, new phenomena began to surface that did not fit into the existing model (e.g., Young's double-slit experiment, photoelectric effect). As a result of these anomalies, new models and theories of light were developed and accepted at various stages, including wave theory, electromagnetic theory, quantum electrodynamics, and general relativity.

In conclusion, the relationship between scientific explanation and argumentation, and how these are scientific practices that scientists use to generate knowledge, can be analyzed and described by the metalanguage of premise, claim, phenomenon (or outcome), data, reasoning, and warrant. Historians, philosophers, and sociologists of science use such metalanguage to describe the nature of science in their work. In the same way, this metalanguage will help teachers and students discuss and subsequently gain a better understanding of scientific practices. This chapter shows how this can be achieved through the incorporation of metalanguage in classroom discourse.

Summary

Learning how science works and the nature of scientific practices is currently an important aspect of learning science. Scientific practice is not about doing science as a kind of "hands-on" experience (Osborne, 2019), but it is ultimately a "language-intensive" engagement that "requires students to participate in class-room science discourse" (NRC, 2014, p. 3). However, in order for teachers and students to participate in such discourse, they will need a metalanguage to talk about the function and structure of language in various scientific practices. Learning the metalanguage of scientific practices in the form of their underlying genres is therefore valuable for students to describe and reflect on scientific practices, and consequently understand their connection to the epistemological underpinning of scientific knowledge.

Toward this objective, I have shown several discourse strategies focusing on the metalanguage of premise-reasoning-outcome (PRO) and claim-data-warrant (CDW), and discussed how they are applied to explanation construction and argumentation in classroom discourse. These discourse strategies require both teachers and students to explicitly learn a shared metalanguage so that they can use it to talk about their own language in constructing explanations and arguments. Among the various genres and practices in science, scientific explanation and argumentation are extremely important not only because they present the most challenge to students, but they also form a tight relationship in the production, evaluation, legitimization, and communication of scientific knowledge.

Thus far, we have only focused on scientific explanations and arguments within one semiotic mode (i.e., verbal). However, the use of visual representations, such as diagrams, sketches, and graphs, is also an integral part of constructing explanations and engaging in argumentation. In addition, the New London Group (1996, p. 77) definition of metalanguage as "language for talking about language, *images*, texts, and meaning-making interactions" also includes the application of metalanguage in analyzing and describing images. Thus, in the next two chapters, we will extend the discussion on scientific practice and metalanguage to incorporate the use of visual representations in scientific explanation and argumentation.

Notes

1 The metalanguage of scientific explanation clearly involves more than premise, reasoning, and outcome, and includes words like principle, law, model, theory, reasoning, cause-effect, sequence, observation, and phenomenon. The selection of premise, reasoning, and outcome is a deliberate choice to simplify the metalanguage for science teachers and students, as well as having an easy acronym (i.e., PRO) to facilitate recall. In addition, these terms are sufficiently broad to subsume most terms; for instance, (a) premise is a broad category that encompasses principle, law, model, theory, and definition, (b) reasoning encompasses deduction, cause-effect, and sequence, and (c) outcome encompasses observation and phenomenon.

2 In an instructional context, many premises in school science explanations use "law-like" rules, which can be further explained by a more fundamental rule. For example, to explain floating and sinking, an acceptable explanation in middle school involves the general rule that the buoyancy of an object is determined by its density relative to its surrounding fluid. However, this "relative density rule" can itself be explained by Archimedes' Principle, which is only taught in high school or university. Thus, what can be accepted as a relevant premise to an explanation must also be considered in relation to the learners' prior knowledge and the curriculum standard at various grade levels.

7

USING DISCOURSE TO

Coordinate Multimodal Translation of Representations

Classroom discourse involves more than the use of verbal language to facilitate talk, interaction, and meaning-making. From Chapter 3 to 6, we have only focused on verbal language, while temporarily putting aside non-verbal representations. The reason is mainly because it is easier to understand some of the key ideas in this book by focusing first on verbal language and subsequently expand those ideas to incorporate other modes. To this end, Chapters 7 and 8 will now expand the discussion we have thus far on discourse patterns and discourse strategies to include multimodal representations.

As explained in Chapter 2, the theory behind multimodality is social semiotics. From this theoretical perspective, any communication system (e.g., speech, writing, image, gesture, mathematics) is a semiotic resource that has been culturally shaped by a community to meet three meaning-making functions – ideational, interpersonal, and textual. Although each semiotic mode has its unique meaning-making affordance, the way we make meaning with every mode is more or less similar; through the "pattern of pattern" realization of meaning from a text to its contexts. For example, just as we make meaning of a written text based on the pattern of its words, we also analyze the meaning of a diagram based on an equivalent pattern of its visual elements. With this theorization, we can examine how teachers and students use images, gestures, and other modes in almost the same way we have done in previous chapters for verbal language.

In Chapters 7 and 8, we will examine the last of the five discourse patterns in this book – multimodal pattern – focusing on how discourse shapes and is shaped by the coordinated use of multimodal representations. This multimodal pattern is further divided into two patterns according to how classroom events operate at two different timescales. The first pattern operates at a macro-event level of classroom activities and focuses on the translation of representations across a teaching sequence. This *multimodal translation pattern* will be the focus of this chapter. In addition, we will focus on the associated discourse strategies in relation to the use of student-generated representations. The second pattern, called *multimodal integration pattern*, operates at the micro-event level of discursive actions where representations across different semiotic modes are used to make meaning. This will be the focus of Chapter 8.

Multimodality in Science

Before we begin, it is important to clarify what a representation is and how it is different from a semiotic mode and an instructional artifact. I will also clarify how a representation is made by drawing from at least one of the five major semiotic modes used in science classroom discourse – verbal, visual, mathematical, gestural, and material, and discuss the affordance for each of these modes.

Mode and Representation

The terms *mode* and *representation* are closely related, but they mean different things in this book. As defined in Chapter 2, a mode is a semiotic system that consists of culturally shaped signs or objects (e.g., sounds, alphabets, logograms) that function as a resource for making meaning. In the history of human civilization, numerous sign systems have been adapted and developed into a cultural resource to help us make and communicate meanings within a specific community. For instance, there are thousands of spoken languages and dialects that have been developed around the world based on the variety of sound made by the human voice, as well as hundreds of sign languages that are developed through the use of hand signals by various deaf communities. All these modes function as semiotic resources with three meaning-making functions: ideational, interpersonal, and textual (Halliday, 1978).

A representation, on the other hand, is a particular form and instance of expression that is drawn from one or more semiotic modes. It is a generalization of the term "text" which was defined in Chapter 2 as an instance of language use. A spoken word, a written statement, a scientific symbol, a picture, or a graph can become a representation if it fulfills three characteristics. First, a representation must have a material form that interacts with our senses (mainly through sight, auditory, and touch). Depending on this materiality, a representation can be classified as an inscription which has a permanent record (e.g., written word, drawing, graph) or ephemeral which is more spontaneous and non-permanent (e.g., utterance, gesture, bodily movement), unless it is recorded on audiovisual equipment.

The first characteristic of representation (on its materiality) is derived from the word "representation" as a concrete noun. However, the same word is also a nominalized term from the verb "represent;" thus it denotes the second characteristic of representation, which is the act and process of representing something by someone to someone else or oneself (Tang, 2013b). In this regard, a representation is a process with a potential to mean something to someone (Peirce, 1986). It is important to stress the word *potential* because meaning is not inherently built into or "carried" in a representation. Rather, meaning must always be actively made as a process by a person interpreting within a particular context.

Third, the meaning potential of a representation is related to the mode from which it is drawn from. For instance, the symbol π is really just three line stokes arranged in a particular way. It does not have any meaning built into it. Neither does it have any meaning to someone without a knowledge of mathematics. This is another way of saying that the meaning of π is interpreted based upon the system of mathematics as a semiotic mode, which was developed over centuries and is currently used by and taught to millions of people. The same applies to all alphabetic words or logogram characters from any writing system (e.g., English, Chinese). Recall in Chapter 4, I gave an example to show that almost all English words in both their phonetic and written form have little resemblance to the actual thing they represent. Words like "fish" can be replaced by any arbitrary labels (e.g., Jeeta, Flo) as long as the semantic connections to the word remain intact. In other words, the meaning of the word (as a representation) is related to the semantic system of language (as a mode) from which it is drawn.

Strictly speaking, most representations are multimodal. What this means is that a representation is made by combining two or more modes. For example, a diagram contains not only lines, shapes, and colors which are drawn from a visual mode, but also symbols and words frequently as well. An oral speech and written text are predominantly a linguistic mode, but they also employ audio elements (e.g., volume, pace, tone, emphasis) and visual elements (font, heading, paragraph, alignment) respectively to stress certain interpersonal or organizational meanings. More elaborated representations used in science instruction such as videos, animations, models, role-plays, textbooks, and worksheets are made from a more complex combination of modes.

One of the challenges in defining a representation is the wide range of "grain size" it can form, ranging from a single sound, word, or line to a fully developed speech, page, or picture (Tang & Moje, 2010). Most researchers and educators in science education tend to see representation as an inscription that is deliberately designed to express or relate to a particular idea or concept. For example, a representation of energy conservation can be described in a written text, illustrated in a diagram or chart, demonstrated through a pendulum, or expressed in a mathematical equation. To resolve any potential confusion in nomenclature, I will call these inscriptions designed with a specific content in mind as *instructional artifacts* and reserve the term *representation* as a more general sign that involves the process of semiosis in all meaning-making.

Classification of Modes

Researchers have generally classified the various semiotic modes used in science classrooms into five major families (Van Rooy & Chan, 2017). Examples of semiotic modes and their corresponding representations belonging to each of these families are shown in Table 7.1.

The verbal-linguistic mode consists of all the thousands of natural languages that are spoken and written today. Each of these languages is a linguistic system that comprises a network of sounds and symbols (letter or character) that can be strung together to form complex meanings in the form of words, clauses, and sentences. We have explored in previous chapters how such complex meanings are made through the semantic and genre patterns generated by verbal language. Strictly speaking, many researchers regard speech and writing as two different modes even though they draw from the same verbal language (e.g., English). This is due to their distinct materiality which shapes their unique affordances: temporal sequence of sound versus visual organization of words (Kress, Jewitt, Ogborn, & Tsatsarelis, 2001). Nevertheless, we can consider them under the same family that draws from a common linguistic system.

The visual-graphical mode comprises an aggregated system of meaning potential that arises from every instance of image use within a culture. As discussed in Chapter 2, there is a system of image use, similar to a system of verbal language, that shapes how images are drawn, composed, and interpreted according to conventions specific to a culture. This system has historically evolved from the way a particular culture has used images to serve various communicative purposes (Kress & van Leeuwen, 2006). In a previous study (Tang, 2013a), I have argued that

TABLE 7.1 Classification of semiotic modes and representations in science classroom discourse

Families of modes	Examples of modes	Examples of representation where the mode is dominant[1]
Verbal-linguistic	Oral and written English, Spanish, Chinese etc.	Written text, oral explanation
Visual-graphical	Scientific images, diagrams, and graphs	Schematic diagram, photograph, computer-generated image, animation (moving images)
Mathematical-symbolic	Algebraic expression, scientific symbols	Chemical equation, formula, vector
Gestural-kinesthetic	Hand movement, body movement, sign language	Gesture, role-play
Material-operational	Laboratory apparatus, 3D models, Braille	A ball-and-stick model of H_2O, Physical setup for a demonstration or experiment

---

[1] Although most representations are multimodal, we can often identify the dominant mode where the representation is drawn from.

every discipline (e.g., art, architecture, science, engineering) has historically developed unique visual systems to make specific ideational, interpersonal, and textual meanings required in the discipline. In science, free-body, circuit, atomic, and anatomy diagrams are some examples of visual images that were developed and used by scientists, which over time, form an aggregated visual system that expands the meaning-making potential of this resource within the scientific community.

The mathematical-symbolic mode comprises a mathematical and scientific notation system that is used to represent quantitative ideas related to science. This system includes mathematical expressions of scientific ideas, algebraic symbols, chemical equations, and computational algorithms. In a study on mathematical classroom discourse, O'Halloran (2000) revealed how several algebraic symbols, such as "+" and "=," function as operative verbs with similar grammatical properties that are found in verbal language. In addition, when an equation is used to express known relationships in science, numerous scientific symbols, such as PE_f and KE_f, function as the specialised entities (analogous to technical vocabulary) while the algebraic operations (e.g., +, =) constitute the grammar of the mathematical expression (Tang, Tan, & Yeo, 2011).

The gestural-kinesthetic mode consists of the use of hand and body movement in the meaning-making process. Researchers have generally documented a number of gesture types – deictic, iconic, metaphoric, emphatic, and emblems (Crowder, 1996; McNeill, 1992). Deictic are pointing gestures that direct an audience's attention to an object, location, or participant in the concurrent speech. Iconic are gestures that resemble a concrete object or movement that is being uttered. Metaphoric are gestures that represent more abstract ideas, actions, or relationships. Emphatic or beats are rhythmic motion or a staccato strike that marks emphasis to an important point in a conversation. Emblems are hand gestures with specific meaning that are well understood in social context, with sign languages as the most developed form of emblems.

Lastly, the material-operational mode comprises the resources and actions that deal with the manipulation of material objects or realia, in what has been commonly called hands-on and tactile activity. Material objects in science classrooms commonly include scientific apparatus and laboratory samples used during practical work (e.g., microscope, test-tubes, chemical solutions, dissection specimen). They also include purposefully designed tools for education purposes (e.g., anatomical model, ball-and-stick model set) or everyday objects that are used to illustrate scientific concepts or phenomena; for instance, a teacher may drop a pencil to demonstrate the effect of gravity. Material objects by themselves do not make up a semiotic mode; it is only when they are being manipulated that they become a resource for making meaning in science. As such, material-operational mode often co-occurs with gestural-kinesthetic and verbal-linguistic mode. Currently, not much research has been done on this particular semiotic mode.

Semiotic Affordance

Each semiotic mode has different advantages and limitations in making meaning compared to other modes. We say that each mode has a unique semiotic affordance or meaning-making potential. For a verbal-linguistic mode, the affordance largely relies on its categorical system of words to make typological meaning or "meaning-by-kind" (Lemke, 2003). For instance, to verbally represent a physical object, we can choose a wide range of words or even coin new words to describe its various attributes (round, soft, bright, gaseous, poisonous, metallic, conductible etc.). As explained in Chapter 2, every choice of word categorizes the object into some kind of grouping in relation to other contrasting words (soft vs. hard, metallic vs. non-metallic). Thus, in a sentence that describes an object, every verb and adjective narrows the object's attributes by categorizing what it is in relation to what it is not, until we gain a sufficient interpretation of its meaning. This is the affordance of spoken and written words to help us make precise qualitative meanings through the categorical system of a verbal-linguistic mode.

However, the verbal-linguistic mode is not adapted to making topological meaning or "meaning-by-degree" (Lemke, 2003) in terms of, for example, *how* round or *how* bright an object is. Most verbal languages have developed adverbs such as *little*, *very*, and *extremely* to give a rough sense of the scale or intensity of a word. However, the extent to indicate the continuous range of the scale or intensity is rather limited. This is where the visual-graphical mode is more suitable due to its spatial and quantitative affordance. For instance, the object's shape can be visually represented more effectively than describing the degree of its "roundness," and its brightness can also be represented on a continuous scale or using different colors when compared to other objects' brightness. Thus, the affordance of a visual-graphical mode tends to center on a continuous variation around spatial and quantitative meanings.

The affordance of the mathematical-symbolic mode rests somewhere in the middle between the two ends of typological and topological meanings (Lemke, 2003). Like written language, mathematical formulations and expressions are generated from a categorical and relational system of signs (e.g., +, x, =) with certain rules or syntax. At the same time, they also denote values that are continuous within the topology of numbers. For example, the work–energy theorem in its algebraic form $- KE_{initial} + PE_{initial} + W_{external} = KE_{final} + PE_{final} -$ is both a typological relationship among the various abstract entities (different kinds of energy) and, at the same time, a topological relationship among the numerical values of the entities that obey the law of conservation of energy. Thus, implied in this mathematical equation is both a causal meaning (e.g., KE change to PE) as well as a quantitative meaning (e.g., KE + PE = constant). As shown in a study by Tang, Tan, and Yeo (2011), this dual typological-topological relationship found in many scientific equations present many challenges for students who are learning science at the secondary school level.

Lastly, the gestural-kinesthetic mode has a unique affordance that combines both spatial and dynamic features. Like the visual-graphical mode, gestures are useful in representing meaning on a continuous scale such as how big, small, long, or short. At the same time, gestures are also often used to indicate temporal pace such as how fast or slow in a particular action. For example, to show the vibration of an object, a teacher may use his fist to simulate the vibration (as an iconic gesture). The movement of the fist can simultaneously represent the frequency of vibration (seen by how rapid the fist moves) and its amplitude (by how far the fist moves). No other semiotic mode can simultaneously combine these spatial and dynamic affordances, with the exception of animation which is a type of moving image (Burn, 2013). But unlike animation, the gestural-kinesthetic mode is much more spontaneous to use as a semiotic resource.

Multimodal Translation Patterns

In this and the next chapter, I introduce two different discourse patterns that are relevant to the use and coordination of representations in science classrooms. These patterns are multimodal translation and multimodal integration patterns. These two discourse patterns are connected by how an *instructional artifact* is used in different timescales of classroom events; ranging from a short interactional exchange that lasts seconds to a series of activities in a teaching sequence that lasts from minutes to hours. (See Lemke, 2002, for a theoretical overview of timescale.) As defined earlier, an instructional artifact is an inscription that is designed to symbolize, communicate, or illustrate a specific idea or concept. Common examples in science classrooms include a passage of written text, diagram, graph, and demonstration kit. Instructional artifacts are ubiquitous in textbooks and other curriculum materials, but they are also frequently made by teachers in the form of presentation slides, whiteboard texts, and worksheets. Instructional artifacts can also be generated by students (e.g., written work, drawing, lab report, physical model).

Multimodal translation focuses on how an instructional artifact is translated or re-represented from one mode to another (e.g., words to diagram to graph) over the course of a lesson or lesson unit (Roth & Tobin, 1997). For instance, when the lesson activity shifts from a demonstration to individual seatwork, there is often a translation of ideas from a predominantly material-operational mode into a written and visual mode. By contrast, multimodal integration examines how various parts within an instructional artifact or across multiple artifacts are combined in classroom interaction to make meaning, usually within a short discursive moment of time (Kress, Jewitt, Ogborn, & Tsatsarelis, 2001). This occurs when teachers or students are trying to make sense of some phenomena using a part of the instructional artifact. To do so, they often need to coordinate their talk, gaze, and gestures with the instructional artifact. For example, a teacher may point at a particular circle in a diagram and say this is a carbon atom. At this moment, the meaning of the circle as representing a carbon atom was made discursively by the integration of speech, gesture, and drawing.

These two discourse patterns are always mutually reinforcing. Multimodal translation at a longer timescale both supports and constrains the kind of multimodal meanings that can be made at a shorter timescale. Conversely, the meaning-making occurring through a multimodal integration at a shorter timescale builds up to the kind of multimodal translation that is possible at a longer timescale. Thus, to understand how representations are used to support the teaching and learning of science, it is necessary to examine both discourse patterns occurring at the two different timescales (Tang, Delgado, & Moje, 2014). Figure 7.1 illustrates the relationship between these two discourse patterns.

In multimodal translation, many educators tend to take for granted or underestimate the translation of representations across a teaching sequence as they assume there is a "natural" connection between the observed phenomenon and all the inscriptions that represent it. This is a common ontological belief held by many people where a scientific inscription "mirrors the world as it 'really' is" (Roth & Tobin, 1997, p. 1076). For example, a free-body diagram, table, or graph of a rolling ball (see Figure 7.1) is regarded as a natural representation of the actual phenomenon. And because these representations are part of nature, there is a fallacy that the translation of an observed phenomenon into various "standard" scientific inscriptions will be natural and straightforward for many students.

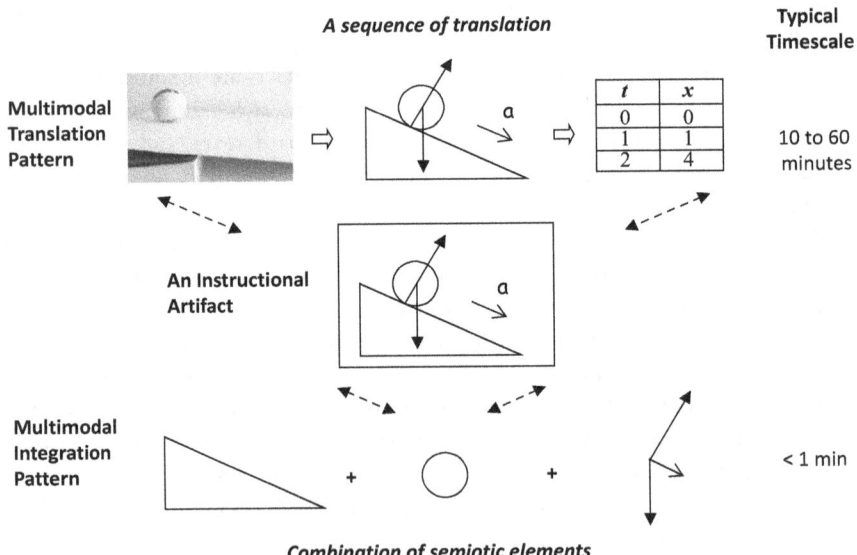

FIGURE 7.1 Relationship between multimodal translation and multimodal integration

However, as we have discussed earlier, representations and their meanings are *made* by people within a particular social context. The reason why many scientists and science educators would regard the translation process of scientific inscriptions as natural is because they have been conditioned to see it that way, instead of seeing it as an integral part of scientific practice. Latour and Woolgar (1979) ethnographic study of scientists (elaborated in Chapter 2) provides a glimpse of how the translation of representations is made as part of scientific practice. In the laboratory, scientists produce a cascade of inscriptions that selectively and systematically transform some features of a phenomenon or data into a network of evidence to support their claims.

Multimodal translation should therefore be regarded as a specific form of scientific practice. I have previously highlighted, in Chapter 6, four of the eight scientific practices in NGSS that are language-intensive. The last practice in "obtaining, evaluating, and communicating information" is related to multimodal translation. In the CCSS literacy for science, this scientific practice is defined as "translating information from one form to another" (NRC, 2014, p. 11). An example of a literacy standard for this scientific practice for Grade 9–10 students would be, "translate quantitative or technical information expressed in words in a text into visual form (e.g., a table or chart) and translate information expressed visually or mathematically (e.g., in an equation) into words" (NRC, 2014, p. 11). These literacy standards tell us that multimodal translation is not a natural process that is unproblematic to students (Roth & Tobin, 1997). Instead, it is a sociocultural way of using language that both teachers and students must pay attention to.

Multimodal Translation Discourse Strategies

The use of multiple representations is a ubiquitous part of science classroom discourse. These representations can be found in all kinds of curriculum materials, including textbooks, lecture notes, worksheets, videos, simulations, and demonstration apparatus. Many teachers have developed implicit discourse strategies on how to use representations as part of their science instruction. These strategies can range from something as simple as showing a video to engage student interest to more complex juxtaposition of multiple representations to illustrate a scientific concept. Aligned with the main theme of this book, this chapter will focus on discourse strategies that involve a more conscious and deliberate action on the part of the teachers and students in their interaction with representations. Three of such strategies will be discussed in this chapter and they are student-generated representation, concrete-pictorial-abstract translation, and dialogic-authoritative transition.

Student-generated Representation

Student-generated representation is increasingly gaining traction as a viable strategy for students to actively engage with representation construction instead of passively learning from pre-existing instructional artifacts made by educators (Prain & Tytler, 2012). Traditionally, instructional artifacts in science classrooms tend to be designed by educators to present content matter to the students. Some common examples include graphs, tables, schematic diagrams, mathematical equations, and demonstration kits. Student involvement with these artifacts tend to be minimal. It may include observing their teachers using a table and translating it to a graph, copying these artifacts drawn by the teacher, or applying them in solving word problems. In recent years, aligning with a constructivist view of teaching, there is now more emphasis for students to create their own representations in order to promote scientific reasoning and knowledge construction. Student-generated representation reflects a changing emphasis from learning from representations toward learning with representations (Tang, Delgado, & Moje, 2014; Tippett, 2016)

There is a variety of instructional artifacts teachers can elect to support student-generated representation. The most common artifact created by students is a diagram, which centers on a visual-graphical mode. Although diagrams are commonly used in science classrooms, a "drawing-to-learn" approach is not the same as copying or interpreting an existing diagram drawn or shown by the teacher. Instead, there is a greater focus on creating a visual model that elaborates or explains the underlying scientific process or phenomenon (Quillin & Thomas, 2015). Logistically, student-generated drawing is easy to enact in the classrooms as it only requires pencils and papers. More advanced technological approaches involving the creation of digital media such as digital stories, animations, and videos are gradually gaining traction (Hoban, Nielsen, & Shepherd, 2013). In addition, student-generated representation can also involve the material-operational and gestural-kinesthetic modes; for example, getting students to create physical models or use their own bodies to model natural phenomena and processes in a role-play (Hubber, Tytler, & Haslam, 2010; McSharry & Jones, 2000).

The purpose of a student-generated representation approach is to promote active learning of a new topic rather than a summative assessment to test what students have mastered at the end of a lesson. Thus, it is important to see the artifacts generated by the students not as final by-products of their learning, but as a multimodal translation process that is co-occurring with classroom interaction. In this regard, what students are essentially doing is translating their experience and understanding from one semiotic mode to another. As every semiotic mode affords different meaning due to its unique affordance, the translation of student representation from one artifact to another provides students with opportunities to build and assess their understanding through different semiotic modes. This is the basis behind the usefulness of student-generated representation as a strategy.

There are two important considerations to make in designing and enacting a student-generated representation approach. The first consideration is how to translate the artifacts generated by students (and the diverse ideas that are represented in them) toward standardized scientific inscriptions that symbolize the accepted ideas in science over a teaching sequence. The second consideration is how to integrate the activity of student-generated representation with classroom discourse. We will discuss these considerations further in the next two sections.

Concrete-Pictorial-Abstract Translation

With regard to how representations are translated in a teaching sequence, it is important to consider the progression of student ideas from a concrete phase consisting of macroscopic objects to more abstract phase consisting of invisible things and ideas (Park, Chang, Tang, Treagust, & Won, 2020; Tang, 2016b). Bruner (1966) provides a useful lens to examine three different phases of thinking involved with using representations to learn: enactive, iconic, and symbolic. The enactive phase involves concrete thinking about real-world objects or phenomena with the use of hands-on materials and bodily actions. The iconic phase involves pictorial or imagery representations that resemble the concrete situation observed or enacted in the enactive stage. These representations can be drawn explicitly or visualized in one's mind. Finally, the symbolic phase involves written words and symbols that are connected to the objects or phenomena being represented through social conventions.

Bruner's concrete-pictorial-abstract phase can be used to design the sequence of how instructional artifacts are translated in a guided inquiry with student-generated representations. At the beginning of a lesson or lesson unit, enactive and iconic types of representations – drawing from material-operational, gestural-kinesthetic, and visual-graphical modes – are most suitable for students to create or manipulate. For example, in a lesson on solid, liquid, and gas, a popular lesson idea is to ask students to use Lego bricks (each representing an atom) to construct their representations of the molecular arrangement in each state of matter. The concrete and tangible nature of this activity allows students to visualize and experiment with their ideas in a three-dimensional space. After this sensory experience, students are asked to draw the molecular arrangement, which is another representation construction but using a different semiotic mode. These two phases of activity – enactive and iconic – provide more creative spaces for students to generate representations that reflect their own ideas rather than conforming to scientific models of representation.

By contrast, the last phase – symbolic – is a stage of representation based on social convention. This is the stage that requires a consensus among a community concerning the use of symbols to represent collective meanings, such as the length of an arrow to represent the magnitude of a force in classical mechanics or dotted line to represent a weaker force of attraction in molecular chemistry. As such, the symbolic phase is not suited for the generation of students' own representation. Instead, students' representation generated during the earlier enactive and/or iconic phases need to be juxtaposed with and gradually translated into canonical scientific inscriptions during the symbolic phase. A key means of doing this is through the accompanying classroom interaction, which we will examine later.

Figure 7.2 shows two lesson examples that illustrate a multimodal translation pattern in the teaching sequence that followed a concrete-pictorial-abstract approach. The first lesson focused on energy transfer in a wave motion. Its instructional objective was to distinguish and describe the movement of particles in a wave motion as a vibration about a fixed point instead of moving along with the wave's forward propagation. Along Bruner's (1966) phases of thinking, the main activities of this lesson involved the students in: (a) generating a transverse wave using a rope and observing several positions highlighted along the rope, (b) discussing and drawing their observation of the positions on a worksheet, and (c) writing their observation and an explanation as to how energy is transferred in a wave motion without the physical transfer of matter (see Tang, Ho, & Putra, 2016). In the last phase (symbolic), it was not only the mode of writing that was more abstract, but also the increasing use of

Phase of Thinking	Lesson on Wave Motion	Lesson on Surface Smoothness
Enactive	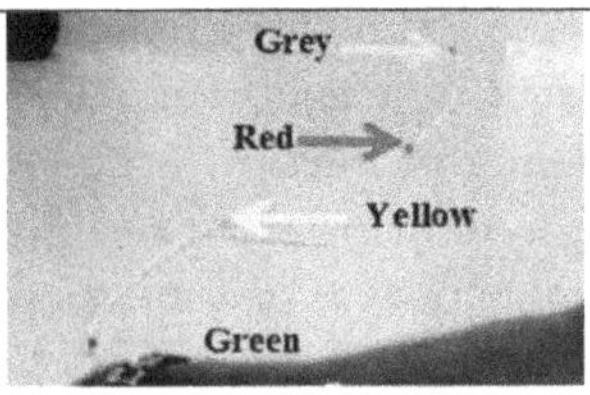	
Iconic	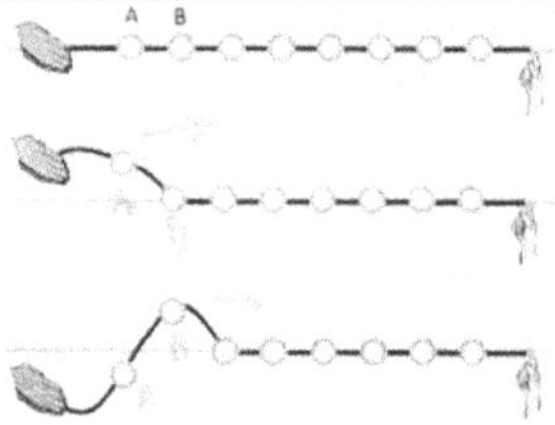	
Symbolic	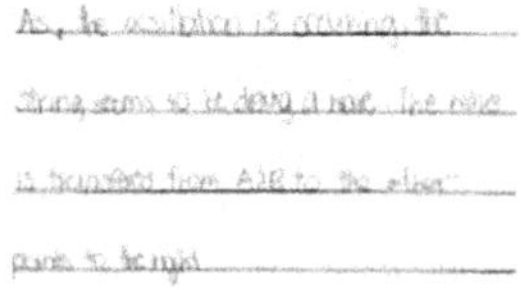	

FIGURE 7.2 Lesson examples that illustrate a concrete-pictorial-abstract multimodal translation

scientific vocabulary and genre (as social conventions) that define this phase as symbolic.

The second lesson focused on the surface property of a material in resisting bacteria build-up. Using salt and different grades of sandpaper as representations of bacteria and various surfaces respectively, the first activity saw the students experimenting with the effect of removing salt residue inside various sandpapers. The second activity then involved the students recording their observation and drawing on an individual worksheet. Finally, the students came together in groups in the third activity to create a group drawing that included more conventionalized drawings and writings (see Tang, Delgado, & Moje, 2014).

In the multimodal translation process shown in these two lesson examples, there was a *dominant* representation that served as the instructional artifact being translated from physical materials and bodily actions (enactive phase), to drawing and visualization (iconic phase), and finally to more elaborated writing (symbolic phase). Throughout this process, it is important to recognize that the interaction at every moment is always multimodal. Thus, what I mean by a dominant representation is the representation that participants orient themselves to in their ongoing interaction as well as the representation that is used to frame and make visible their actions (Goodwin, 2000). For instance, in an iconic phase, when students were drawing their diagrams, they were also frequently talking and gesturing to one another, writing short notes and annotations, and manipulating physical materials at the same time. However, these multimodal actions revolved around the drawing activity in the sense that the act of drawing through a visual-graphical mode was instrumental in shaping or prompting the students to talk, gesture, write, and manipulate objects in a certain way. To understand how this multimodal coordination takes place through a dominant representation will require an understanding of the multimodal integration pattern at a different timescale, and this will be further elaborated in the next chapter.

Dialogic-Authoritative Transition

Begin with Dialogic Interaction

With regard to the second consideration of designing and enacting a student-generated representation approach, we must bear in mind that any learning with representation activity always simultaneously reinforces and is reinforced by the accompanying classroom interaction. Thus, the discursive interaction among teachers and students is crucial in facilitating creation of and learning with representations. In Chapter 3, we discussed that there are two spectrums of interaction in classroom discourse – dialogic (multiple voices) and authoritative (one accepted voice). As student-generated representation aims to encourage students to be creative in generating new ideas instead of conforming to canonical ideas, dialogic interaction is more useful at the beginning in mediating the process of getting students to create their own representations. Conversely, student-generated representations are also multi-voiced (incorporating different perspectives in a non-verbal way) when it co-occurs with a dialogic interaction.

At the beginning of a lesson or lesson unit centering on a key topic or concept, dialogic interaction should be used to elicit inputs from the students before introducing the accepted scientific ideas (Mortimer & Scott, 2003). According to a constructivist teaching approach, this is a useful way to generate interest, activate prior knowledge, and stimulate their thinking process. To this end, discourse strategies that promote IRF interaction and student dialogue as explored in Chapter 3 are particularly useful. For example, we learnt that dialogic questioning is essential in eliciting a range of views in a whole-class or small group discussion. We also learned that the use of a group drawing (as a student-generated representation) and collaborative talk among the students play a mutually supporting role (Park, Chang, & Tang, in preparation).

The following example illustrates a dialogic interaction with student-generated representation at the beginning of a lesson focusing on wave motion. The lesson objective was to explain how particles in a wave vibrate perpendicularly in relation to the forward propagation of the wave. The lesson began with an enactive (hands-on) activity that incorporated material-operational representations. Students in groups were asked to stimulate a number of wave motions using a number of available resources provided in the classroom, such as ropes, long springs, and a ripple tank. They could also use their own bodies as a role-play in the demonstration. Following the enactive activity, they were then asked to draw their observations of the wave motion on a worksheet. The teacher gave this instruction before the activity begun:

EXCERPT 7.1

#	Speaker	Utterance
1	Teacher	Your task is to make a wave motion. You **can be creative** and use any of these materials or your own bodies if you like to **generate the wave**, okay? You work in your groups. When you produce the wave motion, observe what happen and then **draw and describe your observation** ...
2	Teacher	In your drawing, I don't want you to just draw the things you have. I want you to draw the motion you observed in different steps. Can you see there is a **number of boxes** here? Okay, **use them to draw the wave motion**.

Following the teacher's instruction, one of the groups discussed what they would like to do for the demonstration, and the following student dialogue occurred:

EXCERPT 7.2

#	Speaker	Utterance
1	Jen	So what you guys want to do?
2	Leung	Let's do a role-play. Then we can show a Kallang[1] wave
3	Jen	Oh cool. Let's do that . . .
4	Pete	Yah but how do we do that?
5	Leung	There, like that (*performs a standing cheer with two hands raised above the head*). You never see a Kallang wave before?

[1] A human wave named after a local stadium.

In this example that centered on the use of enactive and iconic representations near the beginning of the lesson, there is a high degree of dialogic interaction in the supporting discourse. In the instruction, although it occurred as a teacher talk, there was an open invitation for them to be "creative" (line 1, Excerpt 7.1) and choose any of the available representations for their demonstration. Subsequently, this led to a variety of ideas from the students with regard to how they wanted to generate the wave representation, with one of the groups choosing to do a role-play of the "Kallang wave" (Excerpt 7.2).

Interestingly, in the second part of the teacher's instruction to draw their observation, there was less openness toward what and how to represent in the iconic phase, as compared to creative emphasis in the earlier enactive phase. First, there was only one choice in the mode of representation (i.e., drawing). All of them had to draw their observation in the given worksheet. Second, the worksheet also specified how they should draw the motion according to the sequence of boxes provided (line 2, Excerpt 7.1). Although the students could still choose to represent the observed motion in a variety of ways pictorially, they were also converging their representations by following a certain template. This shows a transition from a dialogic interaction to more authoritative interaction as they moved from enactive to iconic phases of representations.

Transitioning to IRF Interaction

After every group of students had mostly completed their drawings and performed the wave motion to the class, the teacher summarized their representations in terms of their commonalities and differences. In particular, she pointed out that almost all the representations showed some kind of wave motion progressing forward while the particles in the wave were only vibrating up and down. She used this opportunity to raise the key question for the subsequent discussion, "why does the wave move forward when the particles do not?"

EXCERPT 7.3

#	Speaker	Utterance	Move
1	Teacher	Can you explain why the wave moves forward even though the particles are not moving forward? As you said, they are only vibrating up and down.	I
2	Fatimah	Because when one side goes up (*gestures upward motion with her left hand*), the other side comes down (*gestures downward motion with her right hand*). So that makes the wave move forward, I think	R
3	Teacher	What do you mean by one side?	F (probe)
4	Fatimah	The side on the left, then it moves to the right	R
5	Teacher	Okay. Can you show us your diagram and **use it to explain what you mean?**	F (probe)
6	Fatimah	(*Walks to the front and projects her group's drawing on the screen*)	
7	Fatimah	The particle over here (*points at A; see Figure 7.3*) moves up and the one over here (*points at B*) moves down	R
8	Teacher	So the particles in the wave are moving up and down right?	F (paraphrase)
9	Teacher	Okay, but that doesn't explain what is moving forward.	F (reflective toss)
10	Fatimah	The diagram shows that it is moving forward. I don't know how to say this, but this moves when the particles are vibrating	R
11	Teacher	Okay. So you're saying that over time the waveform moves forward right? That is the shape of the wave. We need to be very specific okay to distinguish between the waveform and wave particles. And which one moves forward and which one vibrates up and down.	F (paraphrase)

In Fatimah's initial response to the teacher's question, she was unable to clearly articulate through verbal language the motion of both the wave particles and waveform. She vaguely referred to "one side" and "other side" of the wave (line 2) as well as "the side on the left" moving to the right (line 4). Consequently, the teacher used a number of IRF moves (e.g., probe, paraphrase; see Chapter 3 for elaboration) and Fatimah's diagram to help her be more specific in the description of the movement. First, the teacher made use of what Fatimah had drawn to help her explain what she meant (line 5). This then allowed Fatimah to be more precise in identifying "the particle" on the wave in line 7 through her deictic (pointing) gesture on the diagram. Second, when Fatimah's explanation was still insufficient to account for what was moving forward, this was pointed out in the teacher's reflective toss in line 9. This led Fatimah to use her diagram again in line

10 and gesturing an animated movement to show the sequence of waveforms depicted in Figure 7.3. These two moves made use of the semiotic affordances of a visual representation to distinguish the difference in movement between the wave particles and waveform.

This particular excerpt illustrates that in order to sharpen the students' ideas toward scientific ideas, there is often a coordinated shift between the mode of representation (from enactive to iconic) and classroom interaction (from dialogic to authoritative), as documented more extensively in a study by Tang (2016b). Thus, as the students translated their previous open-ended physical representations (e.g., demonstration, role-play) into a more restricted visual representation, the classroom interaction as facilitated by the teacher also became more authoritative, as exemplified in this episode. For example, the IRF probes and paraphrases were used by the teacher to help Fatimah see from a scientific point of view of distinguishing the specific wave particles as vibrating from the general waveform that moves forward. This was explicitly highlighted in line 11 when the teacher said that "we *need* to be very specific okay to distinguish between the waveform and wave particles." In other words, the use of a visual representation and an IRF classroom interaction

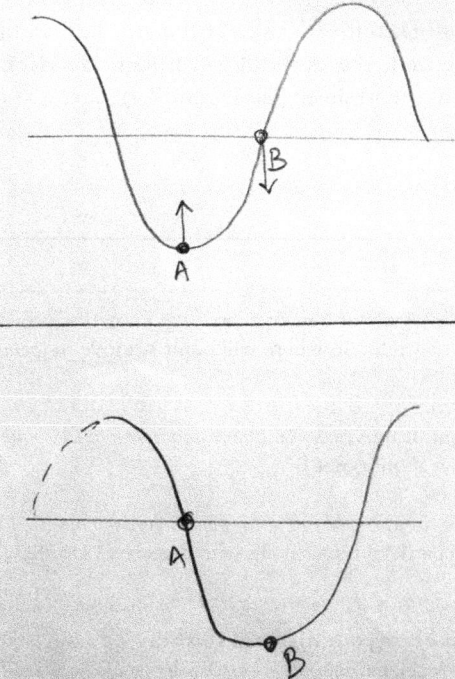

FIGURE 7.3 Drawing of wave motion by a student

goes hand-in-hand in shaping the middle stage of a typical multimodal translation pattern.

Although the student's diagram as a visual representation played a crucial function in this interaction, it is important not to overemphasize its role in relation to the verbal and gestural modes. Reiterating what has been mentioned in the beginning of this chapter, a representation by itself does not contain any inherent meaning. Rather, meaning is always made and interpreted using multiple modes in a situated context. This is how learning occurs through meaning-making *with* all the available representational modes (drawing, talking, gesturing), and not learning *from* (static) representations. As we saw in this interaction, the diagram drawn by Fatimah's group (i.e., Figure 7.3) on its own could not explain the movement of the waveform and its particles. It was only when the diagram was combined with what she said and gestured in lines 7 and 9, in context to the teachers' IRF questioning and prompts, that she was able to explain more articulately how the wave motion occurred. This multimodal *integration* of the various semiotic modes determines how we produce scientific meanings, and it will be further explored in the next chapter.

Rounding up with Authoritative Interaction

To round up the multimodal translation discourse pattern in this lesson on wave motion, we will examine an excerpt toward the end of the lesson where the teacher summarized the key idea through an IRE interaction and a diagram taken from a worksheet (see Figure 7.4):

EXCERPT 7.4

#	Speaker	Utterance	Move
1	Teacher	Okay, after say about 1 second later, we know the waveform travels to the right. So where will point A move to (*points at A*)? Larry?	Initiate
2	Larry	Move down	Response
3	Teacher	That's right. It moves down (*draws an arrow from A pointing down*). Then, how about point B?	Evaluate / Initiate
4	Larry	Up	Response
5	Teacher	Yes. It moves up (*draws an arrow from B pointing up*). So do the particles travel along with the wave? Do they?	Evaluate / Initiate
6	Class	No	Response
7	Teacher	No, right? So, **if you draw your diagram this way**, you can see more clearly how the particles move up or down as the waveform is traveling to the right ...	Evaluate

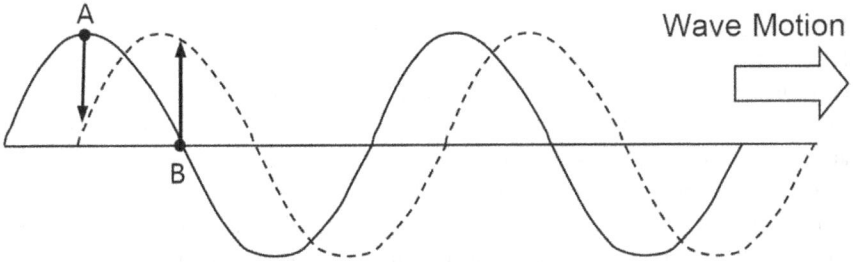

FIGURE 7.4 A canonical representation of wave motion used in Excerpt 7.4

In this final excerpt, the main instructional artifact used to facilitate the discussion was not only a visual representation, it was also a canonical one due to the social conventions associated with such a representation to depict wave motion. The conventions include: (a) drawing another wave (usually in dotted line beside the original wave, so as not to confuse the two waves) occurring at a short moment later, (b) selecting a few points on the wave (usually at the nodes and antinodes), and (c) drawing arrows on the selected points from the original to the subsequent waves. These conventional ways of drawing wave motion are commonly found in classroom talk, textbooks, worksheets, and examination papers around the world. According to Bruner (1966), this symbolic (convention) phase is the final mode of thinking involved in learning with representations. At the same time, this excerpt also shows an authoritative classroom interaction that was used to highlight the salient features in the canonical diagram. Therefore, the use of a canonical representation and an IRE classroom interaction mutually support and reinforce the later stage of a typical multimodal translation pattern.

The introduction of a canonical representation may not have to be delivered from a teacher-centric summary or IRE interaction, as shown in this example. A useful strategy called explicit comparison is to juxtapose a student-generated representation against the canonical representation in order to compare their relative affordances and limitations. This can be carried out prior to, concurrent with, or after this particular excerpt where the teacher summarized the key ideas of the lesson. In addition, a dialogic interaction through whole-class or student discussion can be used to facilitate the explicit comparison. However, at some point, the students will have to be guided to see from a scientific point of view and understand why certain conventions are necessary or beneficial in the canonical representations. The strategy of explicit comparison will be further elaborated in Chapter 8 when I will discuss how to analyze the meanings that are commonly made with representations.

Summary

This chapter has expanded the scope of classroom discourse from a previously dominant verbal mode to include a multimodal assemblage of representations.

Extending from our theoretical understanding based on social semiotics, I elaborated on the difference between a semiotic mode and a representation, as well as the classification and affordances of five major families of semiotic modes found in science classrooms: verbal-linguistic, visual-graphical, mathematical-symbolic, gestural-kinesthetic, and material-operational modes.

An understanding of how a multimodal representation is made is necessary to discuss the last discourse pattern in this book – multimodal pattern. This pattern focuses on how discourse shapes and is shaped by coordinated use of multimodal representations. Multimodal pattern operates at two different timescales. At a macro-event timescale, multimodal translation examines the discourse pattern of translating representations from one semiotic mode to another over a teaching sequence. This pattern of multimodal translation is not a simple or natural process for many students, but is instead a language-intensive scientific practice that must be learned. To do so, a number of multimodal translation strategies such as student-generated representation, concrete-pictorial-abstract translation, and dialogic to authoritative transition were recommended and discussed in this chapter.

As for a micro-event timescale, we will need to examine a multimodal integration pattern. This is a discourse pattern of making meaning at a discursive level through the integration of representations across multiple semiotic modes. This meaning-making pattern will be further examined in the next chapter.

8

USING DISCOURSE TO

Coordinate Multimodal Integration of Representations

In the use and coordination of multimodal representations, there are two different discourse patterns operating at different timescales of classroom events. The first pattern, multimodal translation, was discussed in Chapter 7. This pattern occurs at a macro-level timescale from one representational activity to another (e.g., demonstrating, drawing, writing). Multimodal translation examines how an instructional artifact is re-represented from one semiotic mode to another during classroom discourse, and is useful for understanding and designing a student-generated representation approach that incorporates students' ideas and voices.

This chapter focuses on the second pattern – multimodal integration, which occurs at a micro-level timescale involving the participants' discursive actions during a representational activity. This involves an analysis of how teachers and students coordinate and combine multimodal resources to make meanings that correspond to specific scientific ideas. Multimodal translation and multimodal integration discourse patterns, while operating at different timescales, are always mutually reinforcing. The translation of representations from one activity to another at a longer timescale shapes the kind of meaning that can be made with a representation at a shorter timescale. Conversely, the meaning-making occurring through a multimodal integration at a shorter timescale develops into the pattern of multimodal translation at a longer timescale.

Besides Chapter 7, this chapter also expands previous ideas introduced in Chapter 4, in particular, semantic relationship and thematic pattern. The theoretical ideas in Chapters 4 and 8 are similar in many ways. Each chapter is essentially about how we make ideational meanings through the semiotic resources we use. Chapter 4 limits the discussion on verbal language as it is easier to first communicate the key ideas focusing on one semiotic mode. In this chapter, the same theoretical ideas and analytical tools are generalized to all other semiotic modes. In particular, they will be applied to the visual and gestural modes, as these are the two most common non-verbal modes in science classroom discourse (except at pre-university and tertiary level where mathematical mode becomes prominent). Subsequently, I introduce several discourse strategies that apply an awareness and understanding of multimodal semantic relationships and thematic patterns.

Multimodal Integration Pattern

Semantic Relationship Revisited

In Chapter 4, we learned that the meaning of a word is not contained within the word itself, but is made by its relationships to other words. Anyone who reads these words has to construct their meaning by forming the *semantic relationships* among those words. Semantic relationship provides a general patterned way of describing something through the signs we use. The same semantic relationship can be made by using different words. For instance, "object moves," "object shifts," or "object is moved" are more or less saying the same thing with different grammatical expressions. Broadly speaking, each of these expressions has the same *material-transitivity* semantic relationship, even though there is a subtle difference in the meaning of the verb being used (i.e., moves, shifts, is moved).

Meaning-making with other non-verbal modes works in a similar way, except that instead of words, we use "symbols" that are drawn from a particular semiotic mode (see Chapter 7 for a discussion of mode and representation). In a visual-graphical mode, a drawing for instance comprises several visual elements, such as dots, lines, curves, boxes, circles, and other geometrical shapes. These visual elements constitute the basic meaningful symbolic unit of a diagram, just as words comprise the basic standalone unit[1] in most languages. Thus, similar to the way we have analyzed semantic relationships from the combination of words, we can do the same by examining how different visual elements are joined together. Moreover, just as there are multiple ways of saying roughly the same thing with different words, there are also different ways of drawing the same semantic relationship with different visual elements. As an example, consider Figure 8.1 which shows the drawings from three different students after they observed the movement of a roly-poly toy demonstration. (A roly-poly toy is a round-bottomed doll that will always return to its upright position when toppled due to its distribution of mass.)

Each of the diagrams shown in Figure 8.1 is a unique expression drawn by three different students. Nevertheless, at a more general level, we can recognize that each diagram expresses a similar semantic relationship involving some kind of movement. In particular, each diagram is typical of three common methods of depicting movement in students' drawings (see study by Tang, Won, & Treagust,

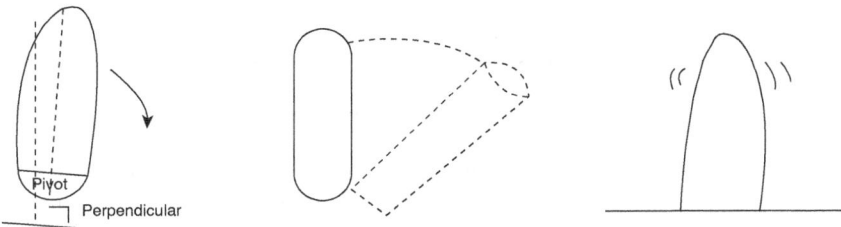

FIGURE 8.1   Three different ways of depicting movement in drawing

2019). The most common and intuitive method is the use of arrows to indicate the direction of the object. The second method is the use of a path line (usually as a dotted line or different color) to trace a previous position of or path taken by the object. The third method is the use of double short wavy lines around the object – a common convention used in comic illustration. The last method may signify the object is moving but it does not indicate the direction and whether the object is traversing, rotating, or vibrating. Although there are nuances among these three methods, we can see there is an overall semantic meaning of *movement* in all the diagrams. This is quite similar to the *material-transitivity* semantic relationship in verbal language. In my earlier example, "object moves" and "object shifts" have the same semantic relationship *linguistically* even though there are subtle differences between moves and shifts.

Semantic Relationships in Visual Mode

I have previously introduced a range of semantic relationships that are constructed with verbal language in Chapter 4. Researchers in social semiotics have developed various analytical frameworks to describe and organize the semantic relationships associated to other non-verbal modes in classroom discourse (e.g., Kress & van Leeuwen, 2006; Lim, 2019; Martinec, 2000; O'Halloran, 2000). It will be beyond the scope of this book to elaborate these frameworks and discuss the semantic relationships for all semiotic modes. In this section, I will discuss the semantic relationships for a visual mode based on a particular framework that I developed from a study of student-generated drawings in science (see Tang, Won, & Treagust, 2019). This framework was adapted from a more general visual framework developed by Kress and van Leeuwen (2006). In addition to the visual mode, I will also briefly discuss how to analyze gestures following the same approach.

The range of semantic relationships commonly made with the use of diagrams is organized into six categories, as shown in Table 8.1. The first three categories – *movement*, *association*, and *spatial* – focus on ideational meaning. The categories of *perspective* and *modality* focus on interpersonal meanings while *connective* focuses on textual meaning.

TABLE 8.1 Classification of semantic relationships in visual diagrams

Category	Semantic Relationship
Movement	Movement – Visual indication of movement, via: • Arrow – using arrow to indicate direction of movement • Path line – using dotted line or color to trace position or path taken • Wavy lines – using short wavy lines around an object No movement – No visual indication of movement
Association	Independent – no visible connection between two visual objects Conjoining – a physical connection between two visual objects, by joining line(s) or adjoining boundary Analytic – a part-whole connection between a larger carrier and its smaller parts or possessions, by joining line(s), adjoining boundary, or inclusion Classifying – a class-subclass connection between two entities, by joining line(s), adjoining boundary, or inclusion
Spatial	Position – location of visual object relative to another, which can be top, bottom, left, right Alignment – alignment of visual object relative to another, which can be parallel, perpendicular, angle, or no alignment Relative size – size of visual object relative to another, which can range from similar to distinct Relative scale – proportion of a visual object relative to another in terms of position and size, which can be exaggerated, realistic, or proportional (drawn to scale)
Perspective	Dimension – portraying visual object in different dimensions, which can be 1D, 2D, or 3D Angle – portraying visual object from a particular point of view, which can be top, side, oblique, mixed (combining top, side, or oblique) Projection – portraying a magnified view of a small part of an object Abstraction – portraying a view that is seen with the naked eye (macroscopic), cannot be seen with the naked eye (microscopic), or based on social conventions (symbolic)
Modality	Formality – depicting credibility of visual object by drawing it as cartoonish (low modality), iconic, or schematic (high modality) Simplicity – depicting credibility of visual object by the use of shades and color to give realism (low modality) or schematic sharp lines (high modality)
Connective	Temporal – indicating passage of time, via: • Numbering – using numbers to indicate sequence of images • Arrow – using arrows to indicate sequence of images • Ordered juxtaposition – using left to right or top to bottom with a sequence (e.g., before, after) Comparative – comparing two or more images side-by-side • Unordered Juxtaposition – using left to right or top to bottom sequence

It will be useful to compare this framework presented in Table 8.1 with the verbal framework that was introduced in Chapter 4 (Table 4.1) as there are many parallel connections. For instance, the categories of *association*, *movement*, and *connective* are somewhat similar to the categories of transitivity-relational, transitivity-material, and logical relation respectively. However, we must be cautious not to stretch the comparison too far due to the unique semiotic affordances from different modes.

As mentioned earlier, in the same way that we can analyze semantic relationships based on the combination of words, we can do the same by examining how different visual objects are joined together. A visual object can usually be identified from its completeness and isolation from other objects in a drawing. The most commonly depicted objects are circles, boxes, and other shapes that form the basic building blocks comprising the drawing. In an analogous way, these shapes function as the noun or noun phrases that represent the things or entities, or "participants" in Halliday's (1994) term, in the experiential world. In most diagrams, it is possible to identify multiple visual objects. Each of these objects can form multiple semantic relationships with other objects, just as there are multiple "participants" in a written text that form semantic relationships with other participants.

Movement

We have earlier discussed the semantic relationship of *movement* that depicts a dynamic action, transition, or unfolding event from a visual object. This action can be identified from the presence of a directional "vector" formed by the depicted objects along a real or imaginary line (Kress & van Leeuwen, 2006). In Figure 8.1, we saw some examples of how students' expressions of movement are typically drawn using three different methods: arrow, path line, and wavy lines. The semantic relationship of movement is parallel to some material processes within the transitivity relation of verbal language, for example "car travels" or "whale swims." However, it cannot represent many other material processes that do not involve any motion; for instance, it will be difficult to draw something that can represent "fire heats" or "bear hibernates." Nevertheless, the affordance of using a diagram to depict movement is that it can also represent spatial direction, orientation, and distance more meaningfully than verbal language.

Association

The next category is what I call *association*, which examines how the visual objects in a diagram are associated to one another through lines, proximity, embedment, and other visual means. Association is comparable to relational process in verbal language, as both of them set up some kind of connections between two separate entities. There are four possible types of association in visual diagram. The first one is *independent* where two objects are not connected by any line or adjoining boundary, thus representing no visible association between them. The second relationship is *conjoining* which signifies a physical connection between two objects, as represented by joining lines or adjoining boundaries. For example, Figure 8.2 shows an independent relationship among the Na^+ and Cl^- circles, while Figure 8.3 shows six conjoining relationships between the touching oxygen and hydrogen circles and two conjoining relationships between the H-O-H compound shapes joined by the dotted lines. In science, conjoining relationships often connote some kind of physical interaction or dependency between two visual objects.

The third type is *analytic* relationship, which is a kind of part-whole relationship between a larger carrier and its smaller parts or possessions (Kress & van Leeuwen, 2006). In Figure 8.4, which is the same as Figure 8.3 with the exception of adding an enclosing box, there are two nested levels of analytic relationship. The first level is the interpretation of each H-O-H compound as being made of two H circles and one O circle. We know this by virtue of the adjoining boundaries among the smaller circles. This corresponds to our knowledge that a H_2O molecule is made up of two hydrogen atoms and one

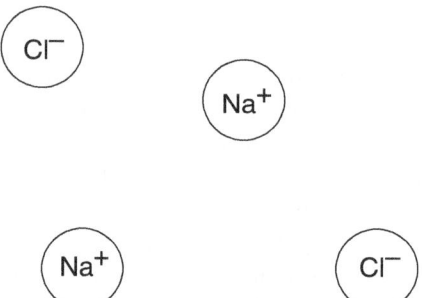

FIGURE 8.2 Example of *independent* relationship

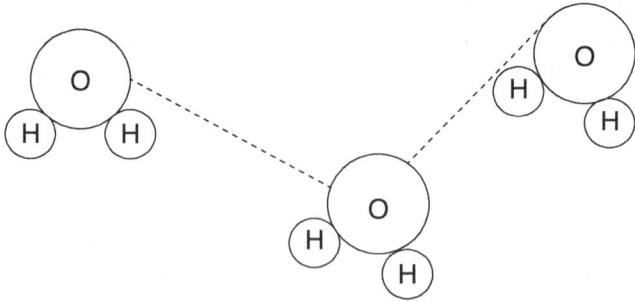

FIGURE 8.3 Example of *conjoining* relationship

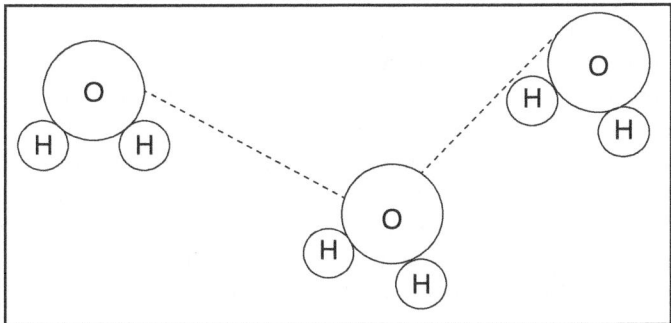

FIGURE 8.4 Example of *analytic* relationship

oxygen atom. The second level is the three H-O-H compounds within an enclosing case, and we know this because they are drawn inside the box. Analytic relationship is similar to the relational processes of *composing* (molecule *is made of* atoms) and *possessive* (molecule *has* atoms) in verbal language, although it is almost impossible to distinguish between composing and possessive in visual diagram.

The last association is *classifying* relationship, which works in a similar way to a subordinate class type of relationship in verbal language. The most common example is a Venn diagram, where the inclusion of a circle within a larger circle or box would represent it as a subset of a larger class. Thus, Figure 8.3 could be interpreted as a classifying relationship if we do not have the context of situation and context of culture. This is why context is important in the interpretation of any text. The role of context was explained in further detail in Chapter 2 through the notions of register and genre.

Besides inclusion of one shape into another, *classifying* relationship can also be made through overlapping boundaries (e.g., intersecting Venn diagrams) or connecting lines (e.g., tree diagram). Classifying relationship is typically applicable to *conceptual organizers* (Hegarty, Carpenter, & Just, 1991), which are abstract diagrams that do not correspond to physical things or connections in the world. This is how we can differentiate between classifying relationship from the earlier conjoining or analytic relationships within the category of association.

Spatial

Spatial relationships highlight how visual objects are related to one another spatially through their *position, alignment, relative size,* and *relative scale.* It is important to note that spatial relationships are relative and they only make sense when we are comparing two or more objects. This is because we can only make judgment on an object's position, alignment, or size only in relation to another object. For example, in Figure 8.5, we can see three visual objects (representing bottles) are *positioned* on the same level from left to right, *aligned* in parallel, and of similar *size and* realistic *scale.* Spatial category is a unique type of relationship that has no parallel comparison with verbal language, unlike association and movement which are comparable to relational and material processes respectively. The closet comparison is perhaps circumstantial relation of location that is embedded within a clause, such as "the bottle is placed *on the floor.*" If we compare with Figure 8.5, we can see how a diagram can provide more affordance to make different aspects of spatial meanings related to the bottle; not just on its position (similar to "on the floor"), but also its alignment and size.

Perspective

Perspective relationships examine the point of view in depicting a visual object. There are four common relationships: *dimension* (one-, two-, three-dimensional), *angle* (top, side, oblique, mixed), *projection,* and *abstraction* (macroscopic, microscopic, symbolic). For example, in Figure 8.6, a student's drawing exhibits a two-dimensional mixed angle that combines both a top and side view of salt residing on a sandpaper. It also shows a projection where the student enlarged the view from a small section of the sandpaper.

Perspective can sometimes be easily confused with the category of spatial. Perspective relationship is a type of interpersonal meaning because it looks at the relationship between the drawer/viewer and the drawn object, in terms of how things or events are portrayed to the audience from a particular point of view. By comparison, spatial relationship is an ideational meaning because it represents the spatial relationships among the objects as they are in the physical world.

Modality

Modality is defined as the credibility or probability of how close our language use resembles reality. It is often reflected in the choice of our words, for instance, "X *could be* Y" reflects more uncertainty (thus lower modality) compared to saying "X *must be* Y" or simply "X *is* Y." In visual image, modality can be identified

FIGURE 8.5 Example of *position, alignment,* and *relative size* relationships

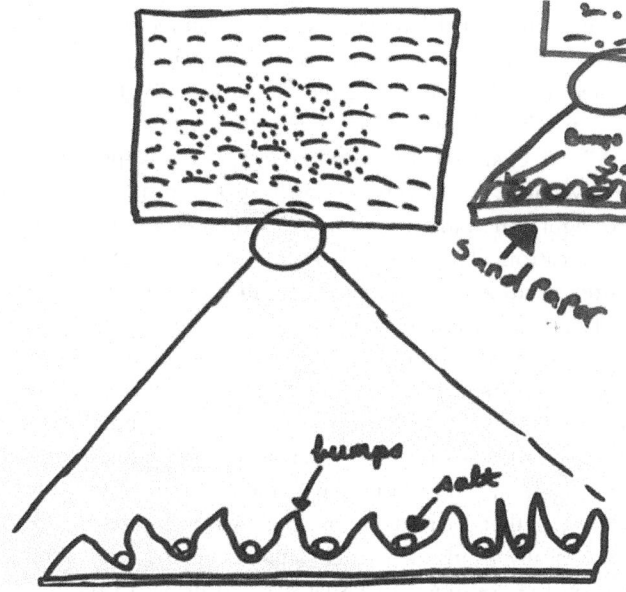

FIGURE 8.6 Example of *dimension* (2D), *angle* (top, side, mixed), and *projection* relationships

according to *formality*, which is the level of details and its perceived credibility, and *simplicity*, which is the style of drawing or imaging using shades, lines, and color. Modality is another aspect of interpersonal meaning as it reflects the reality as constructed by the author and perceived by the viewer.

According to Kress and van Leeuwen (2006), the modality of an image is not universal but depends on the discourse in various disciplines. In mass communication, for example, more credibility is often ascribed to a photograph as compared to an artistic sketch showing the same object or event. Thus, modality in mass communication is related to naturalistic realism. However, in scientific discourse, modality is linked to universality and objectiveness rather than its realism. Thus, black-and-white schematic diagrams have a higher modality because they are deemed to be universal. By contrast, photographs showing the same phenomena or experimental setup have a lower modality because they are seen as localized and limited to a particular laboratory. Modality will come in useful when we discuss the use of images in supporting scientific argumentation, as this scientific practice relies on the credibility of evidence.

Connective

Connective relationships join successive images into a larger coherent narrative, and is a type of textual meaning. They are similar to logical relations that join successive clauses in verbal language through connective words, for example, *and*, *furthermore* (additive); *however*, *while* (comparative); *therefore*, *so* (consequential); *when*, *after* (temporal); *if*, *then* (conditional). In diagrams, successive images are seen when the same visual object is repeated (with some changes). In Figure 8.7 for example, the same object – a roly-poly toy – was drawn three times. What it represents is not a spatial displacement of the object from left to right, but rather a temporal sequence of the same object in the exact spatial location. The student's numbering for each sequence does help with this interpretation, but without this numbering, we will still recognize the overall diagram as a temporal change.

There are two types of visual connectives – temporal and comparative. Temporal sequences can be shown in three ways, by *numbering* (as in Figure 8.7), *arrows*, or *ordered juxtaposition*. The use of arrows in temporal sequence can easily be confused with the semantic relationship of movement, which can be made via an arrow, path line, or short wavy lines. Ordered juxtaposition puts two similar objects together to imply a pre-post action sequence. In most cultures, the order is typically from left to right and top to bottom, although this order also depends on the constraint of visual space available to the drawer. For comparative connective, there is an *unordered juxtaposition* of two or more similar objects for a side-by-side comparison. For example, in Figure 8.8, a student drew the progression of a sound wave in solid and gas to compare their similarities and differences.

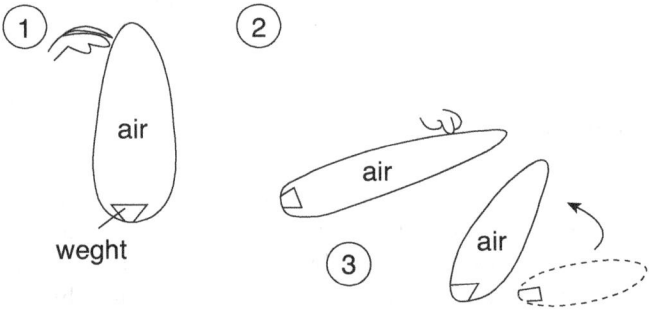

FIGURE 8.7 Example of *temporal connective* relationship

solid: ‖‖∣∣∣∣∣∣∣∣∣ ∣ ∣ ∣∣∖∖∖∣∣∣∣∣∣

gas: ∣∖∖∖ ∣ ∣ ∣∣ ∣ ∣ ∣∖∖∖∖ ∣∣∣ ∣∣ ∣ ∣

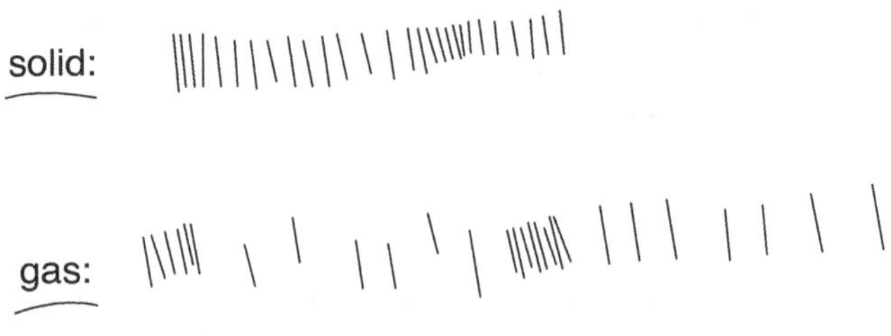

FIGURE 8.8 Example of *comparative connective* relationship

Semantic Relationships in Gestural Mode

This section will give a brief outline of how ideational meanings are made with gestures. As gesture is a huge topic, my treatment here will be highly selective for the purpose of this chapter. For a more comprehensive treatment of gesture, readers can refer to these sources (e.g., Crowder, 1996; Lim, 2019; Martinec, 2000; McNeill, 1992). In Chapter 7, I have described the classification of gestures into the following types, namely, deictic, iconic, metaphoric, emphatic, and emblem. However, this classification only examines the form of gesture, but not the function of their use. This is where I draw on a framework by Martinec (2000) that identifies the functional actions made with gestures according to their realizations of ideational, interpersonal, and textual metafunctions (as inspired by SFL). I will only focus on the ideational metafunction.

Martinec (2000) suggests there are three types of actions that construct ideational meanings in gestures. They are called *presenting action, representing action*, and *indexical action.* Presenting actions are similar to the transitivity processes in language, but now it is the hand gestures rather than words that realize the process. These actions can be real (e.g., pushing a table to demonstrate a force) or imaginary (e.g., mimicking the action of pushing). Following the transitivity system in SFL, presenting actions can be classified into material, behavioural, verbal, and mental. Material process is an action directed at others or physical objects (e.g., pushing a table, scraping a surface), while behavioural, verbal, and mental processes are directed to oneself (Lim, 2019). For science meaning-making, we only need to focus on material process.

Representing actions are gestures with a signifying function to *represent* something (Martinec, 2000). This can include the use of emblems that have conventionalized meaning to the participants (e.g., thumbs up or victory sign). It can also include iconic and metaphoric gestures that represent some kind of movement or shape (e.g., tracing a path, showing extent of size). Typically, these iconic and metaphoric gestures co-occur with speech such that the uttered words will contextualize the meaning of the gesture, and vice versa. Lastly, indexical actions realize ideational meanings in conjunction to the accompanying speech. The most common indexical action is a pointing (deictic) gesture that connects the uttered words to an object or representation pointed by the gesture. Indexical action in gesture is similar to an identifying relational process used in verbal language (See Table 4.1).

Multimodal Integration Pattern

Science meaning-making is often formed through a multimodal coordination among speech, diagrams, and gestures. The essence of this meaning-making can be examined from its multimodal integration pattern, consisting of a network of verbal, visual, and gestural semantic relationships. To illustrate this pattern, I show a particular discursive interaction between a teacher and three middle school students. The context of the lesson – in terms of its multimodal translation pattern – has been described earlier in Chapter 7 (see Figure 7.2). In an earlier representational activity, the students had experimented with salt and various sandpapers that had been established to represent bacteria and surfaces of varying smoothness. In this activity, the students drew on a poster to explain why it was easier to remove salt residue from a fine-grained sandpaper as compared to a coarse-grained sandpaper. In their group poster, they drew three different representations of sandpapers (fine, medium, coarse). Each representation showed a top view and a magnified side view projected from a part of the top view (see Figure 8.6).

When the students had finished drawing and writing on their group poster, the teacher asked them to give their explanation using their diagrams. A critical interaction occurred when two students, Mary and Luke, were frequently pointing at and gesturing over different parts of the poster as they were explaining, as follows:

EXCERPT 8.1

#	Speaker	Verbal Utterances	Video Snapshots
1	Mary	Because when you're scraping,	
2	Mary	you're scraping the top of the bumps	
3	Mary	and you can't get into it	
4	Mary	this one you can get a little bit into the salt	
5	Mary	And this one you can get it	

(Continued)

EXCERPT 8.1 (Cont).

#	Speaker	Verbal Utterances	Video Snapshots
6	Luke	this is the sandpaper right here	
7	Luke	it's trying to get down here	
8	Luke	So basically it's not all the way down	
9	Mary	you can't get it at all	

Using the semantic relationship frameworks for the verbal, visual, and gestural modes, I piece together the thematic pattern of this interaction. Moreover, following the same conventions for drawing thematic pattern in Chapter 4, this thematic pattern is visualized in Figure 8.9. Like all thematic pattern diagrams, the purpose of Figure 8.9 is to show at a glance the "pattern of semantic relationships," so that we can examine it at greater detail. However, unlike previous diagrams that only incorporate verbal utterances (e.g., Figure 4.3), the thematic pattern in Figure 8.9 now integrates the visual and gestural components of the interaction.

At first glance, we can see a pattern of similar and contrasting semantic relationships. Looking at the horizontal connections in the thematic diagram, the material and relational processes are almost the same for each sandpaper. In the verbal mode, we have two main material processes: (i) [scrape − surface] and (ii) [get − salt], repeated three times. In the visual mode, each of the sandpaper images also have the same movement and association relationships (similar to material and relational processes in the verbal mode) in terms of: (i) [salt − no movement] and (ii) [salt − groove − bump] as conjoining.

By contrast, the vertical connections are different, and their differences account for the underlying explanation of this particular thematic pattern. In the verbal mode, there is a difference in scale among "can't get," "get a little bit," and "can get" as well as "not easy," "less easy," and "easy." Rhetorically, these varying degrees of assertions provide the outcome or conclusion of the explanation. As for the basis and reasoning of the explanation, (refer back to Chapter 6 for the structure of a scientific explanation), this would be provided through the visual mode.

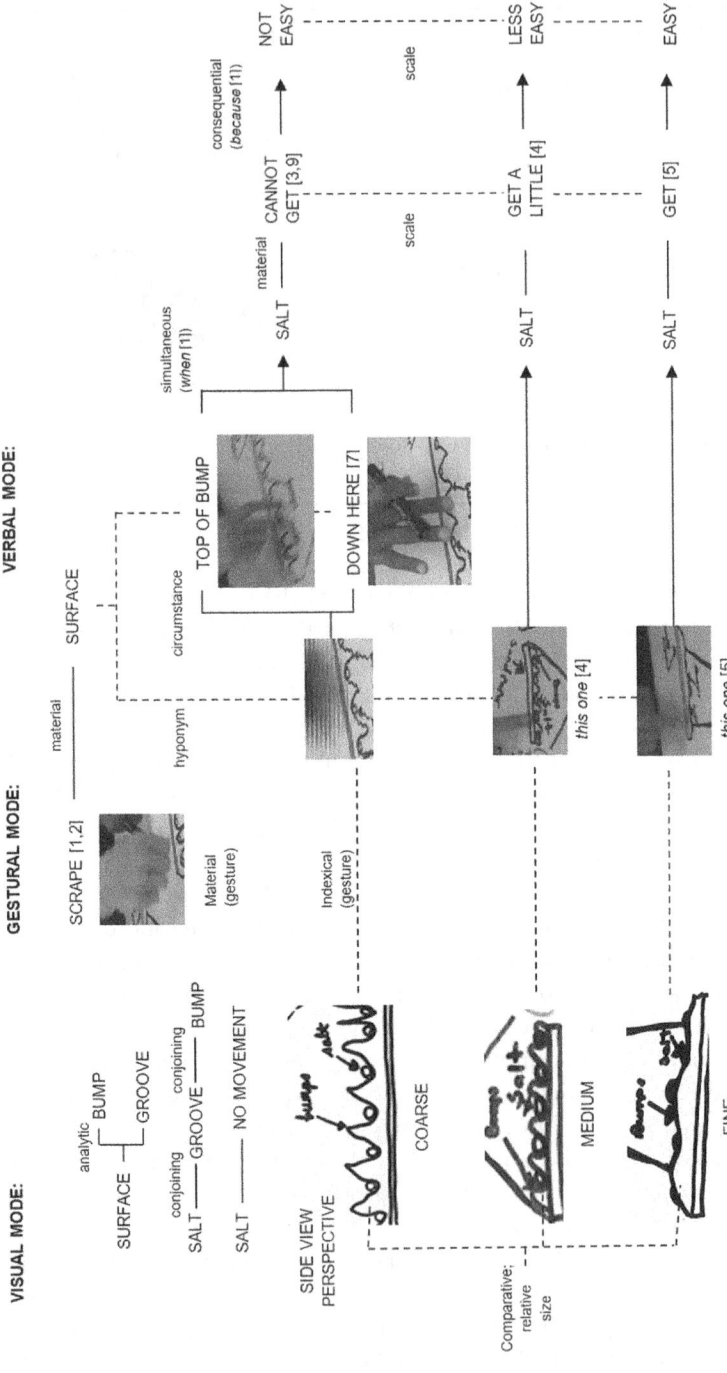

FIGURE 8.9 Multimodal thematic pattern for the interaction

In particular, the main point of difference is the relative size spatial relationship across the three images. I will come back to elaborate how this difference was constructed in the visual images.

Another visible pattern in Figure 8.9 is the complementary roles from each semiotic mode and the nature of the multimodal integration. As mentioned earlier, verbal language provides the two main material processes of "getting the salt" and "scraping the surface" involved in this explanation. It also provides the crucial logical relations as formulated through the conjunctions "when" and "because" uttered by Mary in line 1. Note the arrows in Figure 8.9 that show how the two material processes are joined together by logical relations. However, verbal language alone is insufficient to make a complete meaning of the explanation and this is where the gestural and visual modes are needed.

In this interaction, Mary and Luke's gestures provide two relevant semantic relationships. The first of these is a presenting action (material process) as seen from: (a) Mary imitating a sweeping motion of scraping the surface in line 2 and (b) Luke's finger moving downward to imitate a physical "get into" the salt in line 7. These action processes reinforce the material processes provided by verbal language. The second relationship is the indexical action to complement the frequent use of deictic references in the speech, such as "this one" (lines 4 and 5), "get it" (lines 3, 5, and 9), and "here" (lines 6 and 7). These indexical actions provide the crucial link to the semantic relationships provided by the diagram, which will be examined next. Without these gestures to coordinate between the verbal and visual meanings, it will be near impossible to make sense of this interaction. In the middle of Figure 8.9 that shows the screenshots of gestures, this visually indicates where the semantic gaps arising from the verbal language are, as well as how the various indexical gestural actions fit in.

Lastly, we need to examine the semantic relationships afforded by the visual diagram to see how the thematic pattern is integrated. While the verbal and gestural modes provide the material processes of "scraping surface" and "getting salt" as well as the logical relations (e.g., when, because) in terms of how these processes are related, they communicate very little about how the salt and surface are connected. This gap is complemented by the visual diagram, particularly its association, spatial, and connective relationships. In Figure 8.9, association relationships are shown horizontally for each of the sandpaper images, while spatial and connective relationships are shown vertically across the sandpaper images. In association, salt and surface are visually connected by, first, a conjoining relationship visualizing "salt is next to groove which is next to bump," and second, an analytic relationship visualizing "bump and groove are part of the surface."

These association relationships give a fuller description of the salt and surface, but they do not account for what determines the ease of scraping each sandpaper. For this, we need the spatial relationship, in particular the relative size of the grooves and bumps in relation to the salt. In addition, as the three images are juxtaposed next to one another (not just visually on the poster but also through the student's gesturing at them), the relative sizes are compared across the three sandpapers in order to create a contrast among them. It is only when these contrasts are integrated with the rest of the semantic relationships in verbal, gestural, and visual modes that the whole explanation could be provided by Mary and Luke.

In a later section, the above thematic pattern will be juxtaposed with another one that involved a top view of the sandpapers instead of the side view. As the semantic relationships afforded by both the top and side view perspectives are different, we will see how the resulting explanations are also different.

Multimodal Integration Discourse Strategies

The multimodal integration pattern illustrated earlier highlights how different semiotic modes are put together to make meanings that correspond to a particular scientific idea or concept, such as surface smoothness in the example. Specifically, we looked at meaning-making in terms of the semantic relationships constructed from each semiotic mode as well as how these relationships interact and complement one another across different modes. These semantic relationships are often taken for granted and they remain implicit in science classroom discourse. As such, the discourse strategies for multimodal integration will first involve a conscious awareness of the role of semantic relationships in various semiotic modes. This will then enable more deliberate strategies that will support students in making the necessary semantic relationships that are needed to learn specific scientific concepts.

Critical Multimodal Connection, Not More Representations

One of the popular trends in science education currently is the encouragement to use more representations to enable science teaching and learning. Increasingly, there have been more efforts to use visual and kinesthetic modes to augment or even replace the predominant verbal mode in classroom discourse. This argument is often popularized by the "learning style" theory, which posits that every student has a preferred style that is predisposed to learning from a particular mode. However, recent research has shown that learning style is just a myth without any empirical evidence (Rogowsky, Calhoun, & Tallal, 2015). Unfortunately, according to a study by Prain and Waldrip (2006), many science teachers tend to use multiple representations to promote interest or cater for individual differences in learning styles without considering how the representations can assist or even confuse their students.

The issue here is not whether we should use multiple representations but rather why and how we use them. To address this question, we need to look deeper at the semantic relationships provided by each semiotic mode to understand their meaning-making functions in relation to the overall thematic pattern. Most of the time, there are a few semantic relationships that can only be realized through a particular semiotic mode or combination of modes. I call these relationships the "critical connection" of a scientific concept (Tang, 2011b; Tang, Tan, & Yeo, 2011). They often account for why a particular representation should be used to support the learning of a scientific idea or concept.

A case in point to illustrate the importance of critical connections is the earlier example on surface smoothness, which was based on a study involving 40 sixth- and seventh-grade students (see Tang, Delgado, & Moje, 2014). A critical connection that supported Mary and Luke's talk of surface smoothness was the semantic relationships of relative size and comparative in their visual diagram. However, many students in the same class did not make that connection, and consequently failed to provide a satisfactory explanation of why it was easier to clean a smooth surface (as represented by a fine-grained sandpaper) compared to a rough surface. Just like Mary and Luke, these students had explored the phenomenon using salt and sandpaper (a material-operational mode) and were asked to use a visual mode to support their explanation. However, most of them drew only a top view perspective of the sandpaper, similar to Figure 8.10, instead of a side view perspective. Consequently, the semantic relationship that could be formed using a top view drawing was distinctively different from a side view drawing, even though both of them were visual representations.

In fact, Mary and Luke themselves also used a top view perspective to form their explanation initially. Figure 8.10 was drawn by them before they subsequently went to draw a side view as a magnified projection from the top view (see Figure 8.6). Using the top view perspective, this was how Mary explained the phenomenon as she pointed at the top view drawing of the fine-grained sandpaper:

> The surface is less bumpier, so it's easier to scrape off. The surface has less bumps than coarse.

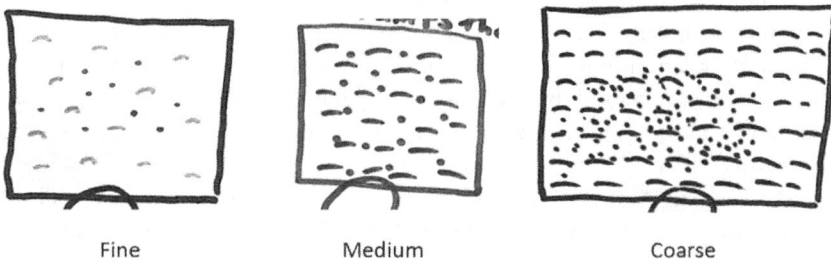

Fine   Medium   Coarse

FIGURE 8.10 Top view perspective of sandpapers

The thematic pattern of Mary's explanation is shown in Figure 8.11. It is representative of six other student groups who gave a similar explanation using the top view drawing of the sandpaper. Comparing this thematic pattern with that of Figure 8.9, we can see what is similar and different. Noticeably, the rhetorical structure formed by the verbal mode is quite similar in terms of the pattern of consequential and scale/antonym relationships. This gave the appearance that both explanations that included a visual drawing were almost equivalent. As a result, the teacher and researchers during the study did not notice both explanations were very different. Only after a post-lesson analysis did they realize that the explanation using the top view was actually misrepresenting the phenomenon (see Tang, Delgado, & Moje, 2014). What is missing in the thematic pattern in Figure 8.11 is the association and spatial relationships in the visual mode that connect the salt, bump, and groove, which were present in Figure 8.9. These are the *critical connections* in this thematic pattern, without which there is no way the students can sufficiently explain the phenomenon.

The above example goes to show that it is not just the use of multiple representations that matter. After all, every group of students used the same set of material and visual representations (e.g., sandpapers, salt, drawing). Instead, it is the semantic relationship that can potentially be formed using those representations that make the crucial difference. In other words, we need to examine and highlight what are the critical connections that form a scientific concept or explanation. Simply giving an array of representations and assuming the students can form the necessary connections does not work (Lemke, 2000). As we discussed in Chapter 7, multimodal translation and multimodal integration are language-intensive scientific practices that must be learned.

A number of studies informed by social semiotics have identified some of the critical connections that are common in several scientific concepts. Table 8.2 highlights some of these studies, which would be useful for science educators teaching these topics.

Explicit Strategies on Multimodal Critical Connection

Understanding the multimodal critical connection behind a scientific concept or explanation is only the first step. The next step is to help students construct the semantic relationships involved in the critical connection as they learn science. Semantic relationships are always implicitly made whenever we talk, write, draw, and gesture in the classroom. Without much guidance, it is often a hit-or-miss affair whether students can form the necessary critical connection from the semantic relationships expressed during classroom discourse. Whenever students face difficulty in their understanding, most teachers will often rephrase what they or other students had said, but this is just another way of repeating the same semantic relationships using slightly different words. Thus,

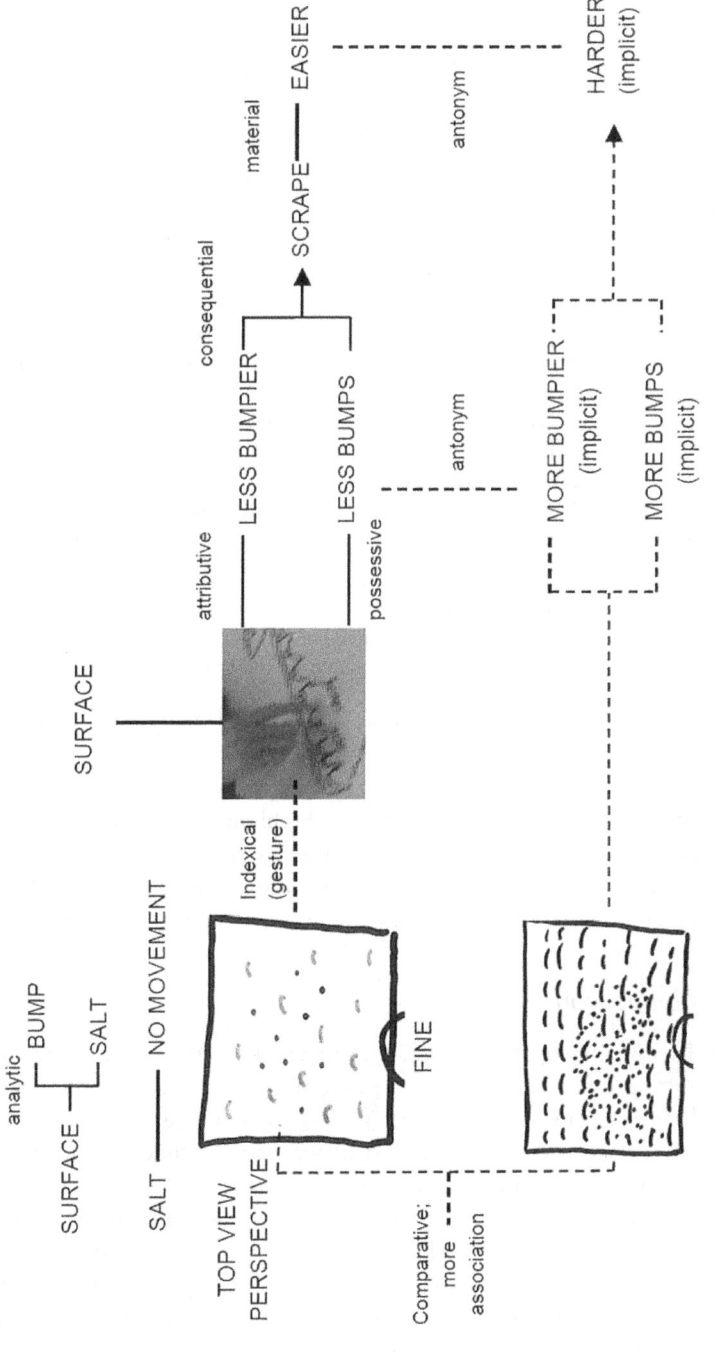

FIGURE 8.11 Thematic pattern with the use of a top view drawing

TABLE 8.2 Critical multimodal connection for various scientific concepts

Concept	Semantic relationships in multiple modes	Reference
Human digestive system	Verbal: physical-material, logical-temporal Visual: association-analytic, connective-temporal Gestural: metaphoric-presenting action (material)	(He & Forey, 2018)
Animal classification	Verbal: taxonomic-hyponym, relational- classifying Visual: association-classifying	(Ge, Unsworth, Wang, & Chang, 2018)
Atomic structure	Verbal: relational-composing Visual: association-analytic Gestural: iconic-representing action (relational)	(Danielsson, 2016)
Force – inertia	Verbal: physical-material, taxonomic-hyponym, logical-temporal Visual: movement, spatial-alignment, connective-temporal Gestural: metaphoric-presenting action (material)	(Williams, Tang, & Won, 2019)
Moment – equilibrium	Verbal: relational-composing, logical-consequential Visual: spatial-alignment, connective-comparative	(Tang, Won, & Treagust, 2019)
Work-energy conservation	Verbal: physical-material, taxonomic-antonym, logical-consequential Visual: spatial-position, connective-temporal Mathematical (based on O'Halloran, 2000): relational, operative	(Tang, Tan, & Yeo, 2011)

we need strategies that are more explicit in highlighting the multimodal critical connection in the semantic relationships. To this end, there is a range of strategies that vary according to its level of explicitness, ranging from explicit comparison, to multimodal analogy, concept-language mapping, and metalanguage.

Explicit Comparison

In explicit comparison, the strategy is to juxtapose two or more different ways of representing for a side-by-side comparison. There is a prerequisite for this strategy before it can be used, and that is to encourage students to generate their own representations. As explored in Chapter 7, student-generated representation allows more student voices into the classroom discourse. The expression of their voices is not limited to a verbal mode, but can be constructed multimodally through non-verbal representations, particularly in drawing. The benefit of generating a number of different representations (especially visual ones) is so that they can be compared more explicitly during classroom discourse. Importantly, the comparison should not be about the form of the representation (e.g., type, look, aesthetic), but should be directed toward their contrasting thematic patterns, in terms of the meaning that can or cannot be made with the use of each representation.

In the sandpaper example, most groups of students drew a top view perspective of the sandpapers compared to a smaller number of groups who drew a side view. At the end of the lesson, each group presented to the class their explanations using their posters. Although explicit comparison was not used at the time during the study due to a lack of time, we can envisage how the comparison could be carried out. Using reflective toss – a discursive technique explored in Chapter 3, the teacher could ask a group of students to compare their drawing with another group who used a different perspective. Besides stating the obvious that one was showing the salt residue from the top while another was showing from the side, the question should prompt them to think about the ideas they could make with each type of diagram. For example, a simple question could be, "what does your diagram show that the other diagram doesn't show?" and vice versa. With some follow-up questions, the teacher could direct the discussion toward the comparison of: (a) the top view showing the *relative number* of bumps and salt and (b) the side view showing the *relative size and depth* of the bumps and grooves. After the comparison, the teacher could ask the students to decide which representation was more useful in supporting their explanation and why.

For visual diagrams, the framework presented in Table 8.1 can be used to guide teachers to ask questions that can facilitate the discussion focusing on meaning instead of form. First, it functions as an interpretative lens to notice and evaluate students' visual representations in terms of a number of categories, notably *movement, association, spatial, perspective,* and *connective.* The identification of these categories helps to discern the subtle differences or gaps that tend to be missed when multiple diagrams are used. As we saw in the sandpaper example, the key difference between the top and side view diagrams, in terms of their meaning, is between association and spatial. The top view is mostly about the inclusion of salt and bumps in a sandpaper, thus leading to an interpretation of the quantity (e.g., more/less bumps). On the other hand, the difference in the side view is the spatial dimension of the sandpaper's attributes, leading to an interpretation of size and depth (e.g., down here, cannot get). Another subtle difference is between movement and (temporal) connective, particularly when it comes to the use of arrows in diagrams. Thus, teachers should always ask students to compare and clarify whether their arrows represent movement in a physical space or denote a sequence in time.

Explicit comparison can also be made across two different semiotic modes that are used to generate representations of the same phenomena. For example, in a sixth-grade lesson on the states of matter, two different student-generated representation activities were used to support the instructional objective of explaining the properties of solid, liquid, and gas. In the first activity, students in small groups used Lego bricks and resealable plastic bags to model the configuration and behavior of molecules in a solid, liquid, and gas inside a container. In the second activity, they drew the molecules that make up a solid, liquid, and gas. Subsequently, the students were asked to use their models to explain a number of questions concerning the shape, volume, fluidity, and compressibility of each state of matter. Prior to these activities, the students had learned about molecules as the smallest unit of matter that retain the matter's properties but they had not seen any molecular model of the three states of matter.

In the next part of the lesson, students were asked to discuss in groups the advantages and limitations for each model. Most students were able to point out the Lego bricks were able to provide a three-dimensional perspective and tactile sensory of the molecular arrangement. However, they also felt the Lego bricks were restrictive in expressing several ideas involving movement and changes. This was where they argued that drawing was better as they could freely draw arrows and other indicators of movement and temporal connective that were necessary for their explanations. Thus, this was an explicit comparison between the affordances of drawing (a visual-graphical mode) and 3D physical models (a material-operational mode).

Multimodal Analogical Reasoning

The use of analogies is not an uncommon strategy in the science classroom (Aubusson, Harrison, & Ritchie, 2006). It is an attractive method for many teachers because it provides a simple way to develop understanding of abstract ideas from concrete references. In the sandpaper example, you may have noticed that the activity of using salt and sandpaper was an analogy that mirrors the interaction between bacteria and surface smoothness. The rough sandpaper models ordinary surfaces with pores that are large enough for bacteria to enter. With the advance of nanotechnology, there are new inventions that can create surface pores smaller than a bacterium. The fine-grained sandpaper models as a nano-smooth surface that can prevent salt from entering it. The analogy is not perfect, but it allows middle school students to think in terms of the relative size of the bacteria and surface pores.

It is useful to think about analogy as a comparable correspondence between two multimodal thematic patterns (Lemke, 1990). An analogy works well when the two patterns have the same semantic relations among the words in each set, i.e., salt and sandpaper versus bacteria and surface. This correspondence is not limited to the use of words within the verbal mode, but also extends to the visual and material modes as well. In fact, the hands-on investigation involving the salt and sandpapers happens to work only because the diameter of an average salt particle sits between the grit size of most sandpapers. Thus, on the basis of this material correspondence, we can build up a similar thematic pattern using the salt and sandpaper to resemble that of bacteria and salt. On learning the salt-sandpaper thematic pattern, which is accessible through visual and tactile experiences, students can then transfer their understanding to a thematic pattern that is less familiar or accessible to concrete experiences. Many common analogies used in science are similar in having this multimodal assemblage; for example, the analogy of water flowing through pipes for electricity, lock and key model for enzyme interaction, and pop-bead model for mitosis and meiosis. In addition, the earlier example of using Lego bricks to model the three states of matter was also an example of analogy.

An awareness of and attention to semantic relationships in multimodal representations can sharpen the use of analogies in the classroom. First, it points us to the "critical multimodal connections" of the underlying concept. In the sandpaper example, it is not surprising that the common semantic relationships between the salt-sandpaper and bacteria-surface thematic patterns are aligned with the critical connections of the concept (i.e., comparative relative size). However, while the critical connections are implied in a good analogy, it is not often the case that the connections are made explicit to the students during classroom discourse. This was a key reason why many students did not make those connections even though they experienced the analogy from the salt and sandpaper experiment. This finding is consistent with other studies that found that teachers seldom talk about the connections when they are using and mentioning the analogy in the classroom (Heywood, 2002; Treagust, Duit, Joslin, & Lindauer, 1992). Therefore, an explicit discussion of the critical connections in the analogy must take place by highlighting the semantic relationships through the classroom discourse.

Concept and Language Mapping

The next strategy also resembles another common instructional strategy, which is concept mapping, developed by Novak, Bob Gowin, and Johansen (1983). This resemblance arises because thematic patterns, when drawn (see Figures 4.3, 8.9, and 8.11), appear to look like concept maps. Although there are some similarities between thematic patterns and concept maps, there are two crucial differences. The first difference is the level of detail. A concept map is a visual representation that shows how the concepts in a discipline are interconnected by "stripping away all text except for concept labels" (Novak, 2010, p. 32). At a general level, it can provide a useful bird's-eye view of the hierarchical knowledge of a domain area. However, it does not account for how a concept itself is assembled through the multimodal representations and discourse used by people (Tang, 2011b). The second difference is the theoretical view of what a concept is. Concept mapping theorizes concept as an abstract idea that resides in the mind, and therefore focuses on an ideal construction of what concepts *should* look like. On the other hand, a thematic pattern theorizes concept as a recurring and repeatable pattern that is constructed through discourse, and therefore focuses on the *actual* connections made among the words, materials, and actions when people speak, write, draw, and act.

Notwithstanding the differences, the benefits of concept mapping and thematic pattern can be combined into an approach that utilizes their relative strengths: the visual organization of concept mapping and the attention to language in thematic pattern. This was the premise that informed the "concept + language" or C+L mapping pioneered by Angel Lin and her colleagues (He & Lin, 2019; Lin & He, 2017). C+L mapping is a pedagogy that includes a suite of teaching materials and activities specifically designed to support language learners in learning science content and a foreign language at the same time in a CLIL classroom. While particularly useful for language learners, some of the instructions are also applicable for non-language-learners as well.

As the term suggests, concept and language mapping integrate both the emphasis of showing the conceptual linkage among the key labels, as well as the semantic relationships among these labels, notably the transitivity, taxonomic, and logical relations. One useful method to integrate them is to juxtapose the thematic pattern with a text passage that gives rise to the pattern. The advantage of this juxtaposition is to facilitate students to read and translate the thematic pattern from the actual text passage, and vice versa. The text passage can be taken from a textbook that the students are reading or adapted by the teacher as lesson handouts. At the same time, the thematic pattern can be initially drawn by the teacher as a form of scaffold, but as the lesson develops, the scaffolding should increasingly fade off to allow the students to co-construct a partial thematic pattern or draw the entire pattern on their own.

An example of a concept and language map is shown in Figure 8.12 where the multimodal thematic pattern (left side of figure) is juxtaposed with a typical text that explains the transformation of mechanical energy in a roller coaster ride (right side of figure). This particular text was taken from a physics education website (Henderson, 2007), and it includes an animation that shows the changing position of the roller coaster in relation to the change of height, speed, KE (kinetic energy), PE (potential energy), and TME (total mechanical energy). In Figure 8.12, I emphasize a number of grammatical features that students should take note of as they read the text passage. Transitivity relations (realized by verbs) are underlined while logical relations (realized by conjunctions) are bold. I also divide the passage (from one long paragraph) into three different temporal stages, according to the spatial-circumstantial relations in the text – "at the top of the hill," "first drop," and "as ride continues." Finally, I highlight using boxes the verb "transformed," which appears three times in the text, before they are "nominalized" (see Chapter 4) into the key concept of the text in the form of the noun phrase "transformation of mechanical energy."

With the multimodal thematic pattern next to the text passage, students can interpret the semantic relationships from the text passage to the thematic pattern,

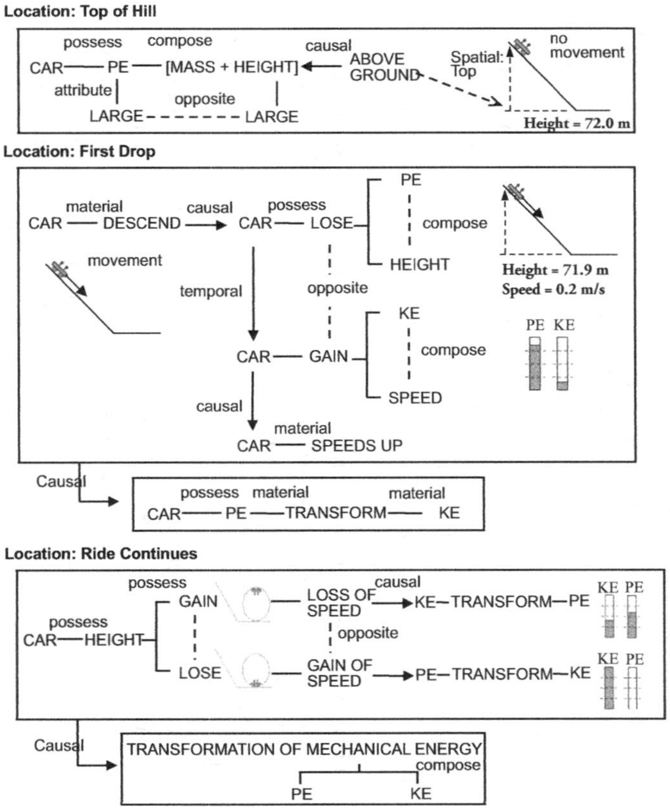

Location: Top of Hill

```
          compose
possess  ┌──────────────┐        causal    ABOVE
CAR── PE──[MASS + HEIGHT]────────────────► GROUND
                │                              ▲
         attribute │ opposite                  │
              LARGE ─ ─ ─ ─ ─ LARGE            │
```

Spatial: Top

no movement

Height = 72.0 m

Location: First Drop

```
         material        causal      possess         PE
CAR── DESCEND ─────────► CAR── LOSE ──── ┌── compose ──┐
                                          │            HEIGHT
  movement      temporal      opposite   │
                                │
                           CAR── GAIN ──── KE
                                          └── compose ──┐
                              causal  material         SPEED
                           CAR── SPEEDS UP
```

Height = 71.9 m
Speed = 0.2 m/s

PE KE

```
Causal │   possess   material
       │ CAR── PE ──TRANSFORM── KE
       │
       │ CAR── PE ──TRANSFORM── KE
```

material

Location: Ride Continues

```
         possess   GAIN
CAR── HEIGHT ──┐ causal
               │ opposite
               LOSE
```

```
LOSS OF     causal
SPEED ──────────► KE──TRANSFORM──PE
  │ opposite
GAIN OF
SPEED ──────────► PE──TRANSFORM──KE
```

KE PE

KE PE

```
Causal │  TRANSFORMATION OF MECHANICAL ENERGY
       │              │ compose
       │          ┌───┴───┐
       │         PE       KE
```

At the top of the hill, the cars _possess_ a large quantity of potential energy. Potential energy – the energy of vertical position – _is dependent_ upon the mass of the object and the height of the object. The car's large quantity of potential energy is **due to** the fact that they _are elevated_ to a large height above the ground.

As the cars _descend_ the _first drop_, they _lose_ much of this potential energy **in accord with** their loss of height. The cars **subsequently** _gain_ kinetic energy. Kinetic energy – the energy of motion – _is dependent_ upon the mass of the object and the speed of the object. The train of coaster cars _speeds up_ **as** they _lose_ height. **Thus,** their original potential energy (due to their large height) _is transformed_ into kinetic energy (revealed by their high speeds).

As the _ride continues_, the train of cars are continuously _losing and gaining_ height. Each gain in height _corresponds_ to the loss of speed (due to speed) _is transformed_ into potential energy (due to height). Each loss in height _corresponds_ to a gain of speed **as** potential energy (due to height) _is transformed_ into kinetic energy (due to speed). This _transformation_ of mechanical energy from the form of potential to the form of kinetic and vice versa is illustrated in the animation below.

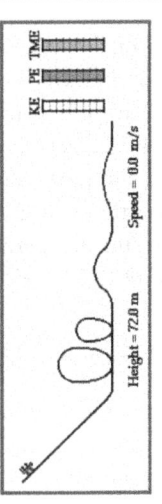

Height = 72.0 m   Speed = 0.0 m/s

KE PE TME

FIGURE 8.12 A multimodal thematic pattern juxtaposed with the multimodal text of the concept

and vice versa. In particular, the thematic pattern visibly shows the critical connections of the underlying concept and how it was derived. First, the critical connection of [PE – TRANSFORM – KE] was built upon the causal consequence from a set of repeating (e.g., lose PE, lose height; gain KE, gain speed) and opposite relationships (e.g., lose height – gain speed; gain KE – lose PE). This critical connection was then repeated and reinforced in the third paragraph "as the ride continues." Notice that the language in this paragraph is denser and more abstract than the previous one; for example, "each gain in height corresponds to the loss of speed" and "continuously losing and gaining height." However, the thematic pattern helps students to see that the semantic relationships are still similar, particularly the semantic relationships that build the [PE – TRANSFORM – KE] and [KE – TRANSFORM – PE].

Metalanguage for Multimodal Discourse

The last explicit discourse strategy is the use of a metalanguage to "talk about the language, images, texts, and meaning-making interactions" (New London Group, 1996, p. 77). The notion of metalanguage was introduced in Chapter 6 as a tool for teachers and students to explicitly discuss the use of language in describing and enacting scientific practices. However, in that chapter, we only examined a verbal metalanguage (e.g., PRO, CDW) and bracketed the role of images and equations that are frequently used in scientific practices. Thus, we need to develop a metalanguage to describe the use of non-verbal representations.

As a start, the categories of various semantic relationships for visual diagrams shown in Table 8.1 can be used by teachers and students to discuss the underlying ideas represented in their drawings. For instance, students can use the metalanguage of *association* (as conjoining, analytic, classifying etc.) to communicate explicitly how their visual objects are connected to one another and explain what those connections represent. Or they can use the metalanguage of *movement* and *connective* to differentiate whether their arrows represent spatial movement or temporal change. This will help teachers and students to pay attention to the multifaceted meanings of their drawing more explicitly.

Moreover, the visual semantic relationships in Table 8.1 can also be used as a framework to develop assessment rubrics or checklists for teachers and students to evaluate their diagrams in relation to the key ideas of specific topics. This can be used both as a formative assessment for students to reflect and continually improve their drawings as well as a summative assessment to evaluate the range of relationships that are represented through their drawings. The purpose of these assessment tools is not to have students learn and reproduce a narrow and standardized way of drawing as students can create alternative ways of representing the same semantic relationships. Instead, the metalanguage in these assessment tools can facilitate an explicit comparison (see previous discussion on this strategy) of the multimodal semantic relationships between students' representations and standardized scientific inscriptions.

In developing a metalanguage for non-verbal representations, it is important to recognize two characteristics of scientific discourse. First, I mentioned earlier that scientific practices are seldom carried out through verbal language alone as they often involve the use of images and equations. Second, the reverse is also true. Scientific inscriptions are rarely produced and interpreted outside any scientific practice (or genre from a discourse perspective; see Chapter 2). This explains why most images used within a particular genre tend to share several characteristics even when the content and topics are different. For example, images used in experimental reports tend to depict a schematic setup of apparatus in terms of their associative connections to one another, while images in explanations tend to highlight changes and processes through the dominant use of arrows and sequential diagrams. Thus, the semantic relationship of *association* is more prominent in experiment report while *movement* and *connective* are more prominent in explanation.

Current research is exploring how to integrate the metalanguage of images into the PRO and CDW metalanguage (discussed in Chapter 6) for explanation construction and argumentation respectively. For instance, there is often a mutual coordination between the reasoning in premise-*reasoning*-outcome and the semantic relationships of *movement* and *connective* in visual images (Tang, 2019a). This is because a scientific explanation revolves around causal and temporal processes of a phenomenon. As for argumentation, visual images are often used as the evidence to support a particular claim. As the persuasion of peers is an integral part of argumentation, the modality of the image in terms of its credibility and universality becomes an important consideration. Thus, there is a mutual coordination between the data in claim-*data*-warrant and the semantic relationship of *modality* in visual images.

Therefore, a metalanguage for describing the combined meaning-making of verbal language and non-verbal representations (especially images) within scientific genres provides a potential useful tool to support science classroom discourse. This metalanguage can be used as both an analytical tool for researchers to analyze the multimodal discourse of science and a pedagogical tool for teachers and students to explicitly discuss how multimodal representations are combined to produce, legitimize, and communicate scientific knowledge. This prospect is still in an early stage of research but with more research in the near future, there will be more pedagogical resources and examples to support science teachers in this area. This was exactly how all the other discourse strategies introduced in this book had developed over the years, from its early experimentation and documentation from small-scale classroom research to its later applications and scaling-up to more science classroom settings around the world.

Summary

This chapter rounds up all the discourse patterns and strategies that are introduced in this book. It continues the discussion on multimodal translation started in Chapter 7 and delves deeper into the teachers' and students' moment-by-moment discursive actions as they coordinate and combine multimodal representations to make meanings. It also broadens the discussion of what constitutes a scientific concept in Chapter 4 and a scientific genre in Chapter 6 by expanding the range of "language use" from a predominant verbal mode to include all the other visual, gestural, mathematical, and material modes.

Specifically, this chapter describes how we make meaning with non-verbal modes based on a similar notion of semantic relationship and the linguistic approach we took in Chapter 4. Instead of looking at the relationship among words, a multimodal analysis entails identifying the smallest meaningful symbolic elements pertinent to the mode and piecing together the semantic relationships among those elements. This is how we analyze meaning-making with visual diagrams and gestures, which are the two most common non-verbal modes used in science classroom discourse. A multimodal integration pattern can therefore be examined by firstly, identifying the semantic relationships from each mode that are used in an interaction, and secondly, joining these relationships together to see how they connect and complement one another across different modes.

Multimodal integration pattern helps us to understand how we combine different semiotic modes to make meanings that correspond to a particular scientific concept or genre. It highlights the multimodal semantic relationships that we often assume or take for granted that students are able to make on their own. In addition, every scientific concept and genre has a "critical multimodal connection" that comprises several semantic relationships that must be realized through a particular combination of modes. This critical connection explains why a particular representation must be used to support the learning of a scientific concept or genre, and deconstructs the myth of introducing the use of multiple representations according to learning style theory.

The critical multimodal connection of a concept or genre can be emphasized and reinforced through a number of discourse strategies at varying level of explicitness. Explicit comparison highlights to students the meaning-making affordances from two representations (from the same mode or across different modes) when they are juxtaposed and discussed. Multimodal analogical reasoning and concept-language mapping expand two common instructional strategies in science teaching to help teachers and students talk about the multimodal connections as they make analogies and concept maps. Finally, the use of metalanguage is the most explicit among all the discourse strategies as it empowers teachers and students to develop a specific language to discuss how multimodal representations are integrated to produce the various genres in scientific discourse.

Note

1 Technically, a morpheme is the smallest unit of language. However, there are many morphemes that appear as parts of words (e.g., ex-, un-) and they do not stand alone as an independent unit of meaning. By contrast, every word consists of one or more morphemes, and it can be used independently. Semantic analysis focuses mostly on the relationships among words rather than morphemes.

9

CONCLUSION

The central theme in this book is the use of discourse strategies in the science classroom, with classroom being broadly defined as a cultural space where multiple discourses intersect. Every science teacher – consciously or unconsciously – uses a range of discourse strategies to construct the teaching and learning of science in five crucial ways: (a) establishing classroom interaction, (b) building thematic content, (c) developing scientific narrative, (d) enacting scientific practices, and (e) coordinating multimodal representations. Discourse strategies, as methods that all humans use in all walks of life to engage in a meaningful interaction, are therefore a fundamental resource that every science teacher possesses. It is not an exaggeration to say that discourse strategy is the most basic toolkit of the teaching profession.

However, most of the discourse strategies used in the classrooms are unconscious or implicit such that teachers are not aware of their existence and usage in their instruction. This creates two issues. First, many strategies are not used appropriately in alignment with the purpose of instruction; for example, asking too many authoritative questions or not using metadiscourse sufficiently to connect students to the lesson. These tend to be the issues faced by inexperienced teachers. On the other hand, many experienced science teachers have developed the art of using a repertoire of effective discourse strategies to enhance their science teaching. But as these strategies are often implicit, it becomes difficult to identify them for beginning teachers to learn. Thus, there is a need to transform the tacit knowledge and skills from teachers into a codified knowledge that is visible for reflection and communication. This is essentially one of the key goals of this book, which is to make teachers' implicit discourse strategies *explicit*. An essential approach to make a strategy explicit is to identify its connection to an underlying discourse pattern of science classroom discourse.

Only by making the strategies explicit, can we then discuss their nature and application, modify and teach them to other teachers and also students, and ultimately have control over our discourse to help us achieve scientific literacy. In this concluding chapter, I will elaborate the nature of explicit discourse strategies, provide some pedagogical recommendations, and discuss the long-term vision of learning discourse strategies.

Nature of Discourse Strategies

Association to Discourse and Its Patterns

The discourse strategies introduced in this book are connected to a particular discourse pattern of how people participate in social activities and make meaning from and of their interactions. These social activities and meaning-making practices occur in recurring patterns in the way we use language in various situations. In other words, discourse strategies are not just techniques that facilitate classroom instruction, but they are fundamentally tools to help students make meanings of what is going on, what they are learning, how to participate, and why they are doing whatever they are doing. Consequently, the application of discourse strategies requires a broader sociocultural view and a more thoughtful consideration of the underlying discourse patterns. For example, questioning is not simply a strategy to elicit a straightforward answer, but an important resource that shapes the interaction patterns between teachers and students and among the students. Using multiple representations is also not a strategy to cater to students' diverse interests, but it is related to a social practice of making meaning through the integration of different semiotic modes.

In this book, I define discourse as a social pattern in the use of language that shapes and is shaped by the way we think, act, and make meanings. I also organize the book according to five major types of discourse patterns: interaction pattern, thematic pattern, narrative pattern, genre pattern, and multimodal pattern, which is further divided into multimodal translation and integration patterns. Each of these discourse patterns was extensively discussed in every chapter from Chapter 3 to 8. In addition, I highlight 20 exemplary discourse strategies that are developed from an understanding and application of the discourse patterns. I have called these strategies: *dialogic questioning, follow-up moves, collaborative questioning, critical semantic connection, unpacking abstraction, jumbled sequencing, narrative framing, heteroglossic projection, epistemological marker, PRO metalanguage, CDW metalanguage, explanation construction, argumentation, student-generated representation, concrete-pictorial-abstract translation, dialogic-authoritative transition, explicit comparison, multimodal analogical reasoning,* and *concept-language mapping.* These discourse strategies and their connections to each of the discourse patterns is shown in Figure 9.1. Many of the ideas for these strategies were adapted or inspired from other researchers, whose seminal works are indicated in Figure 9.1, as well as my own research projects which are further described in the Appendix.

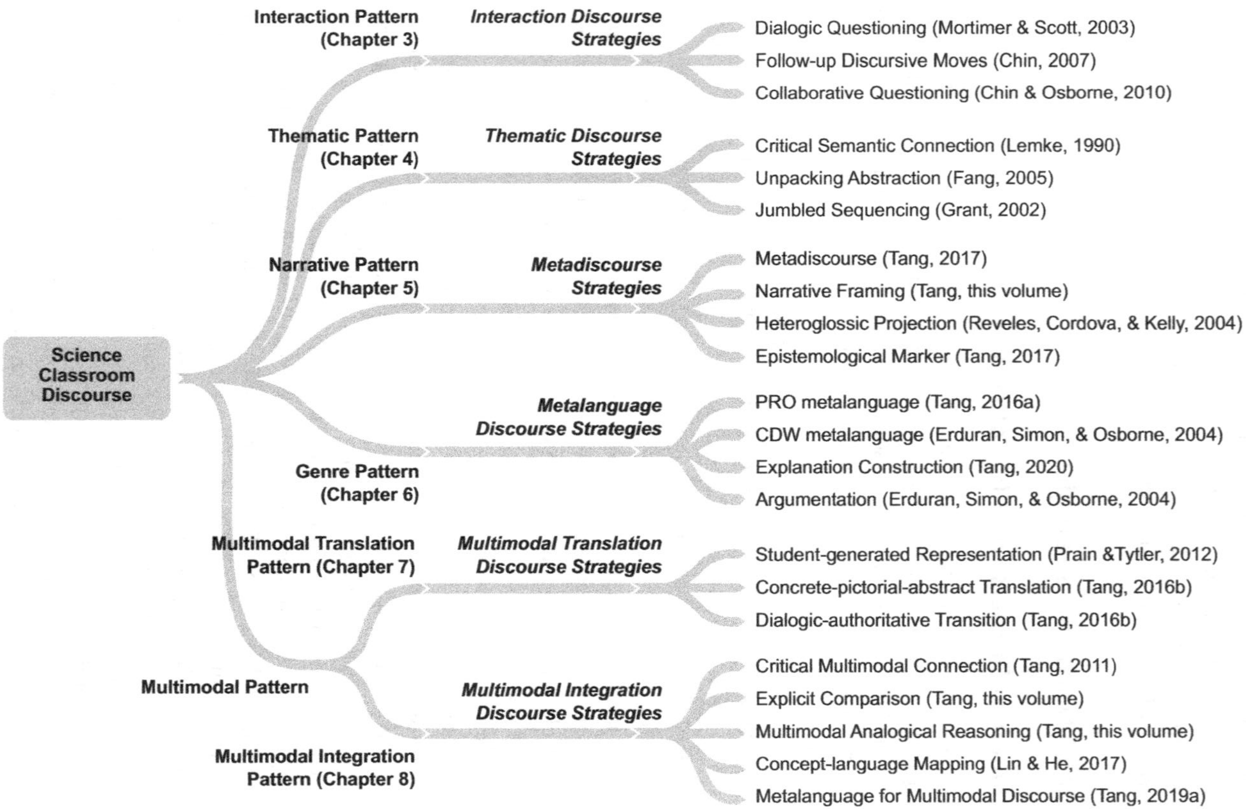

FIGURE 9.1 Exemplary discourse strategies and their relation to discourse patterns

Explicitness and Naming of Discourse Strategies

In developing and naming these 20 discourse strategies, I made a distinction between implicit and explicit strategies. While acknowledging in Chapter 1 that this distinction is not binary, it has nevertheless been useful in this book for two reasons. First, it draws our attention to many implicit strategies that teachers are already doing unconsciously and it highlights what more needs to be done or changed to improve their teaching practice. For example, the use of close-ended questions to build and maintain IRE interaction is a discourse strategy, according to our definition from Gumperz (1982). This strategy is now a routinized practice in almost every classroom, largely due to the apprenticeship of observation (Lortie, 1975) that every teacher went through as a student. By making these implicit routines *explicit* and giving them names (e.g., IRE interaction), this brings into our consciousness the appropriate use of such practices and facilitates reflection on how to apply them. The same is true for the discourse strategies that most teachers intuitively use to make semantic relationships (Chapters 4 and 8) and talk about their own discourse (metadiscourse, Chapter 5) when they talk science. By making the semantic relationships and our metadiscourse more explicit and naming them for both teachers and students, this will better accentuate their use and application in supporting students to make scientific meanings.

Second, the distinction draws our attention to notable strategies that a small handful of exemplary teachers are doing which are distinctive from the more common strategies that we know of. For these exemplary teachers, these strategies might be implicit to them as the strategies have become second nature over the years of their teaching experience. Dialogic questioning, follow-up moves, narrative framing, heteroglossic projection, and concrete-pictorial-abstract translation are some examples of such strategies. For the purpose of professional development, it is valuable to highlight these notable strategies and make them explicit, so that they can be shared and adapted by more teachers. Then, there are also strategies that are already quite explicit in nature because they are developed or emphasized by researchers as pedagogical interventions. Examples of such interventions include PRO metalanguage (Tang, 2016a), argumentation (Erduran, Simon, & Osborne, 2004), student-generated representation (Prain & Tytler, 2012), and concept-language mapping (Lin & He, 2017). The very nature of the research intervention entails that teachers are conscious of their deliberate use and application to achieve certain goals associated with the interventions.

Therefore, what I have called an explicit form of discourse strategy varies according to individuals and situations. It is important to recognize our effort to make a strategy explicit is only temporarily for the purpose of communication and teacher training. Once teachers have learned this new strategy (via explicit methods), the goal is for them to practice using and adapting it until it becomes second nature (implicit) to them. In other words, there is an iterative cycle to this implicit-explicit nature. First, we need to make implicit discourse strategies explicit so that they can be observed, codified, discussed, communicated, shared, adapted, and eventually learned. Once the explicit strategies are learned or relearned, the goal is to transform them into an implicit practice through an extended period of classroom application, evaluation, and reflection.

Finally, one of the most practical steps to make an implicit discourse strategy explicit is to name it. Applying what we have learned from Vygotsky (1986), our thought and consciousness of some phenomena are always mediated by a language. Linking a strategy to a specific name not only provides a reference to what has been discussed in this book, but it also provides a shared label that facilitates further communication and reflection. This was the rationale for how I have named the 20 discourse strategies in this book (see Figure 9.1).

Practical Considerations

The 20 discourse strategies in this book should not be taken as isolated and off-the-shelf techniques that can be applied immediately in the classroom. Instead, a more holistic consideration of the associated discourse patterns and the ideas discussed in the respective chapters is needed. In light of this cautionary remark, I summarize some of the key ideas and recommendations from this book in Table 9.1.

Discourse Strategies for Whom and for What?

In concluding this book, it seems appropriate to discuss the purpose of learning discourse strategies; who gets to benefit from them and for what? I reiterate from what has been said previously that the discourse strategies in this book benefit all science students. This is because discourse strategies provide access to the knowledge and ways of knowing in a particular discourse, which are the products of specific social institutions and cultural practices (Kelly, 2007). To understand how different groups of people benefit from learning discourse strategies, we need to first identify who tends to get access and marginalized in science education to the knowledge and practices of science. There are three groups of people that come to mind with regard to access and marginalization.

The first group is language learners who are still learning the language of instruction while they are learning science at the same time. Language learners face an additional language barrier when they have to learn science in a foreign language, and thus more attention and strategies need to be directed to address their specific learning needs (Lo & Lin, 2019). Due to international migration, there will be more language learners around the world, so this issue will increasingly become more prominent in the future. At the same time, there are also indigenous populations in many countries who do not speak the

TABLE 9.1 Summary of discourse patterns, strategies, and recommendations

Discourse Pattern	Discourse Strategies and Recommendations
Interaction Pattern: *A pattern of conversational exchange between two or more people in terms of their normative roles, expectations, and participation structures*	• Promote an appropriate mix of teacher talk, IRE and IRF interactions, and student dialogue according to different requirements in a lesson. • Use more dialogic questioning for teachers who tend to be more authoritative. Ask more open-ended questions involving "why" or "how." Invite multiple perspectives. • Shift IRE triadic exchange to an IRF chain by using a range of discursive moves, such as extend, probe, paraphrase, reflective toss, and constructive challenge. • Foster student dialogue by providing guidance and modeling on how to ask collaborative questions. Use IRF discursive moves as prompts for student questioning. Pair student questioning with various cooperative strategies.
Thematic Pattern: *A pattern of semantic relationships among words that are used to describe a thematic content*	• Examine the critical semantic connection from a key passage that will be read by or taught to students. • Take note of the missing or assumed semantic relationships that tend to occur in classroom talk. • Clarify or highlight the semantic relationships made by students frequently before going further into the content development. • Unpack abstraction that arises from multiple semantic relationships that are condensed into a technical word or phrase, by "reverse nominalizing" its noun phrase back to its material verb. • Use jumbled sequencing to support students in making logical relations in a scientific explanation.
Narrative Pattern: *An organizational pattern of opening, closing, framing, sequencing, and linking classroom activities as well as an evaluative pattern characterized by a preferred ideological stance toward the content*	• Use metadiscourse judiciously to talk about the content as it develops through classroom talk. • Narrative framing – Use text and activity connectives to assist students in following the narrative structure in a lesson by setting context, marking boundaries, linking cohesive and temporal segments, and signaling topic shifts. Use attitude, epistemology, interpretative markers to highlight the stance required to interpret the content matter. Assist students to connect to the stance by discussing alternative stances and why they are needed. • Promote the talk on science inquiry by using heteroglossic projection for students to adopt the voice of a scientist and incorporate it into their own voice. • Use epistemology markers more frequently for both teachers and students to signal the evidential basis of their knowledge and talk about the epistemic source of scientific knowledge.

(Continued)

TABLE 9.1 (Cont).

Discourse Pattern	Discourse Strategies and Recommendations
Scientific Genre Pattern: *A cultural pattern and set of conventions that are associated with and partly realize the way scientists carry out their practices with language*	• Teach students the structure of scientific explanation and argument using metalanguage, such as PRO (premise-reasoning-outcome) and CDW (claim-data-warrant). • Embed the teaching of metalanguage in authentic science activities involving puzzling phenomena, investigations, or data collection. • Develop non-sequential reasoning in explanation construction with the use of PRO metalanguage as navigational markers. • Promote students' debate through competing theories or competing predictions argumentation with the use of CDW metalanguage. • Encourage students to evaluate the validity of an explanation and argument using metalanguage.
Multimodal Translation Pattern: *A pattern of translating or re-representing an instructional artifact from one semiotic mode to another over a teaching sequence*	• Create opportunities for learning with representations by focusing on student-generated representation. • Apply Bruner's concrete-pictorial-abstract approach to guide student thinking with representations from concrete to abstract phase. Use enactive and iconic types of representations at the beginning and gradually transitioning to symbolic types. • Use dialogic interaction to elicit and support students' ideas in generating diverse representations. Direct students' ideas toward scientific ideas through a coordinated shift of representations (from enactive to iconic to symbolic) and classroom interaction (from dialogic to authoritative). • Discuss the relative strengths and limitations between student-generated representations and standardized scientific inscriptions.
Multimodal Integration Pattern: *A pattern of integrating the semantic relationships across different semiotic modes that are used to describe a thematic content*	• Examine and identify the critical multimodal connections that form a scientific concept or explanation. • Clarify or highlight the semantic relationships across different modes made by students in relation to the critical multimodal connections. • Juxtapose and contrast the underlying semantic relationships from two or more different ways of representing in an explicit comparison. • Reinforce analogical reasoning by comparing and contrasting similar semantic relationships in two corresponding thematic patterns. • Use concept-language mapping to juxtapose a multimodal thematic pattern next to a text passage (including diagram), showing the conceptual linkage and semantic relationships among key labels in a concept. • Develop and use a metalanguage to talk about the use of drawings in a particular genre, such as a scientific explanation or experimental report.

official language very well. Many of the discourse strategies in this book are applicable to language learners because these strategies often address the discursive, linguistic, semiotic, epistemic, and affective challenges in accessing scientific language (Tang, 2019b).

The second group of students comes from communities with discourses that are not well positioned to interact with scientific discourse. This notion of discourse conflict was briefly discussed in Chapter 2. Much research has shown that children from non-European working class families often participate in home literacy practices that do not align well with mainstream classroom practices (Aikenhead, 2001; Heath, 1983). For instance, many of these children are not used to the kind of IRE and IRF question-and-feedback interaction as compared to middle class children who frequently encounter such interactions during dinner and bed-time conversation (Ochs & Taylor, 1992). Therefore, we need to question our cultural assumptions about how science should be taught and adopt a wider range of discourse strategies from a sociocultural perspective. In particular, discourse strategies that make explicit the normative interaction in the classroom, the ideological meaning-making and narratives of science, and the epistemic practices in the discipline are extremely beneficial. These strategies were covered in this book, but undoubtedly, much more needs to be done to create a more equitable science education.

The last group comprises the rest of the "mainstream" population who are traditionally not viewed as marginalized when it comes to science teaching and learning. For these students, access to scientific discourse is equally as important, albeit for a different reason. The discourse strategies in this book are beneficial to these students as the strategies support numerous aspects of science classroom teaching and learning. No doubt many teachers will use these strategies to improve student learning and achievement, as increasingly mandated by curriculum standards and high-stakes examination around the world. However, there is another important reason why discourse strategies are important to these students, and this has to do with gaining control of a discourse for scientific literacy. In order to gain control of a discourse, one must first understand how the discourse works through the language we used in that discourse. Making discourse patterns and strategies explicit in the classroom thus has the benefit of surfacing several features of scientific language and understanding how those features are used to frame scientific knowledge and practice. This helps students to discern the underlying ideological perspective and social practices involved in scientific discourse and consequently develop a critical literacy of evaluating how our knowledge of the natural world is constructed by and through discourses (Tang, 2011a).

Scientific literacy is generally defined as a person's knowledge and ability to participate in science-related issues and decision-making as a reflective and responsible citizen (OECD, 2016). It involves more than a disciplinary understanding of the theories, concepts, and processes of science, in what has been called Vision I of scientific literacy (Roberts & Bybee, 2014). Scientific literacy also involves a Vision II, which focuses on the human and social dimensions of science in terms of its roles and connections to socio-political issues. Along with Vision II, the long-term goal of learning and using discourse strategies is ultimately to enable people to read scientific texts and take part in scientific discourse. Doing so not only enables them to make better personal decisions in the domain of science and technology, but also to question the motives and power relations among groups of people involved in consensus-building in a modern democracy. This is an important endeavor for *all* students.

As I am writing these last few pages, most places around the world are currently in lockdown due to the spread of the coronavirus disease or COVID-19. This unprecedented crisis has caused widespread panic among many people globally. Suddenly, science-related ideas such as pandemic, "herd immunity," and "flatten the curve" are now everyday buzzwords used by politicians and journalists. Mathematical concepts that most adults learned from schools and have probably forgotten (e.g., exponential growth, logarithmic graph) are now thrust into our consciousness again. The current situation calls for an urgent need for greater scientific literacy; not so much in producing scientists to develop a vaccine or doctors and nurses to treat the sick, but more for *everyone* to do the right thing. This includes following the discourse on what is happening, understanding what is at stake, making responsible everyday decisions, and most importantly in today's post-truth era, knowing who to trust as our source of information and advice during this crisis. Thus, scientific literacy is not a distant vision, but a very immediate concern that impacts all our lives and wellbeing. The purpose of learning scientific discourse, and the discourse strategies that promote it, is therefore to facilitate this vision of scientific literacy.

REFERENCES

Abell, S. K., & Lederman, N. G. (2007). *Handbook of Research on Science Education*. Mahwah, NJ: Lawrence Erlbaum.

Achinstein, P. (1983). *The Nature of Explanation*. New York, NY: Oxford University Press.

Aikenhead, G. (2001). Integrating Western and Aboriginal sciences: cross-cultural science teaching. *Research in Science Education, 31*(3), 337–355. doi:10.1023/A:1013151709605

Anderson, C. (2007). Perspectives on Science Learning. In S. K. Abell & N. G. Lederman (Eds.), *Handbook of Research on Science Education* (pp.3–30). Mahwah, NJ: Lawrence Erlbaum.

Aubusson, P. J., Harrison, A. G., & Ritchie, S. M. (Eds.). (2006). *Metaphor and Analogy in Science Education*. Dordrecht: Springer.

Bakhtin, M. M. (1981). *The Dialogic Imagination: Four Essays*. Austin: University of Texas Press.

Bakhtin, M. M. (1986). *Speech Genres and Other Late Essays* (1st ed.). Austin: University of Texas Press.

Barnes, D., & Todd, F. (1977). *Communication and Learning in Small Groups*. London, England: Routledge & Kegan Paul.

Becker, H. (1954). Vitalizing sociological theory. *American Sociological Review, 19*(4), 377–388. doi:10.2307/2087456

Braaten, M., & Windschitl, M. (2011). Working toward a stronger conceptualization of scientific explanation for science education. *Science Education, 95*, 639–669.

Brown, J. S., Collins, A., & Duguid, P. (1989). Situated cognition and the culture of learning. *Educational Researcher, 18*(1), 32–42.

Bruner, J. S. (1966). *Toward a Theory of Instruction* (Vol. 59). Cambridge, MA: Harvard University Press.

Burn, A. (2013). The Kineikonic Mode: Towards a Multimodal Approach to Moving Image Media. In C. Jewitt (Ed.), *The Routledge Handbook of Multimodal Analysis* (pp.375–385). London: Routledge.

Bybee, R. W., Taylor, J. A., Gardner, A., Van Scotter, P., Powell, J. C., Westbrook, A., & Landes, N. (2006). *The BSCS 5E Instructional Model: Origins and Effectiveness*. Colorado Springs, CO: BSCS.

Chin, C. (2006). Classroom interaction in science: teacher questioning and feedback to students' responses. *International Journal of Science Education, 28*(11), 1315–1346. doi:10.1080/09500690600621100

Chin, C. (2007). Teacher questioning in science classrooms: approaches that stimulate productive thinking. *Journal of Research in Science Teaching, 44*(6), 815–843. doi:10.1002/tea.20171

Chin, C., & Osborne, J. (2010). Students' questions and discursive interaction: their impact on argumentation during collaborative group discussions in science. *Journal of Research in Science Teaching, 47*, 883–908. doi:10.1002/tea.20385

Crawford, B. A. (2000). Embracing the essence of inquiry: new roles for science teachers. *Journal of Research in Science Teaching, 37*(9), 916–937.

Crowder, E. M. (1996). Gestures at work in sense-making science talk. *Journal of the Learning Sciences, 5*(3), 173–208.

Danielsson, K. (2016). Modes and meaning in the classroom – the role of different semiotic resources to convey meaning in science classrooms. *Linguistics and Education, 35*, 88–99. doi:10.1016/j.linged.2016.07.005

Duschl, R. A., & Osborne, J. (2002). Supporting and promoting argumentation discourse in science education. *Studies in Science Education, 38*(1), 39–72.

Erduran, S., Simon, S., & Osborne, J. (2004). TAPping into argumentation: developments in the application of Toulmin's Argument Pattern for studying science discourse. *Science Education, 88*(6), 915–933. doi:10.1002/sce.20012

Erickson, F. (1992). Ethnographic microanalysis of interaction. In M. D. LeCompte, W. Millroy, & J. Preissle (Eds.), *The Handbook Of Qualitative Research In Education* (pp. 201–225). New York: Academic Press.

Erickson, F. (2006). Definition and Analysis of Data from Videotape: Some Research Procedures and Their Rationales. In J. L. Green, G. Camilli, & P. B. Elmore (Eds.), *Handbook of Complementary Methods in Education Research* (pp. 177–191). Abingdon: Routledge.

Fairclough, N. (1992). *Discourse and Social Change*. Cambridge, MA: Polity Press.

Fang, Z. (2005). Scientific literacy: a systemic functional linguistics perspective. *Science Education, 89*(2), 335–347.

Fisher, R. (2002). Shared thinking: metacognitive modelling in the literacy hour. *Reading, 36*(2), 63–67. doi:10.1111/1467-9345.00188

Flavell, J. H. (1979). Metacognition and cognitive monitoring: a new area of cognitive-developmental inquiry. *American Psychologist, 34*, 906–991.

Fontaine, L., Bartlett, T., & O'Grady, G. (2013). *Systemic Functional Linguistics: Exploring Choice*. Cambridge, UK: Cambridge University Press.

Foucault, M. (1972). *The Archaeology of Knowledge*. New York: Pantheon Books.

Friedman, M. (1974). Explanation and scientific understanding. *The Journal of Philosophy, 71*, 5–19.

Garcez, P. M. (2017). Microethnography in the Classroom. In K. A. King, Y.-J. Lai, & S. May (Eds.), *Research Methods in Language and Education* (pp. 435–447). Cham: Springer International Publishing.

Ge, Y.-P., Unsworth, L., Wang, K.-H., & Chang, H.-P. (2018). Image Design for Enhancing Science Learning: Helping Students Build Taxonomic Meanings with

Salient Tree Structure Images. In K. S. Tang & K. Danielssson (Eds.), *Global Developments in Literacy Research for Science Education* (pp. 237–258). Cham, Switzerland: Springer.

Gee, J. P. (2011). *Social Linguistics and Literacies: Ideology in Discourses* (4th ed.). New York: Routledge.

Goffman, E. (1974). *Frame Analysis.* Cambridge, MA: Harvard University Press.

Goodwin, C. (2000). Action and embodiment within situated human interaction. *Journal of Pragmatics, 32*(10), 1489–1522.

Grant, K. (2002). *Supporting Literacy: A Guide for Primary Classroom Assistants.* New York: Routledge.

Gumperz, J. J. (1982). *Discourse Strategies.* Cambridge; England: Cambridge University Press.

Halliday, M. A. K. (1978). *Language as Social Semiotic: The Social Interpretation of Language and Meaning.* London, England: Arnold.

Halliday, M. A. K. (1985). *An Introduction to Functional Grammar.* London, England: Arnold.

Halliday, M. A. K. (1993a). On the Language of Physical Science. In M. A. K. Halliday & J. R. Martin (Eds.), *Writing Science: Literacy and Discursive Power* (pp. 54–68). Pittsburgh: University of Pittsburgh Press.

Halliday, M. A. K. (1993b). Some Grammatical Problems in Scientific English. In M. A. K. Halliday & J. R. Martin (Eds.), *Writing Science: Literacy and Discursive Power* (pp. 69–85). Pittsburgh: University of Pittsburgh Press.

Halliday, M. A. K. (1994). *An Introduction to Functional Grammar* (2nd ed.). London: E. Arnold.

Halliday, M. A. K., & Hasan, R. (1976). *Cohesion in English.* London: Longman.

He, P., & Lin, A. M. Y. (2019). Co-developing science literacy and foreign language literacy through "concept + language mapping". *Journal of Immersion and Content-Based Language Education, 7*(2), 261–288. doi:10.1075/jicb.18033.he

He, Q., & Forey, G. (2018). Meaning-Making in A Secondary Science Classroom: A Systemic Functional Multimodal Discourse Analysis. In K.-S. Tang & K. Danielsson (Eds.), *Global Developments in Literacy Research for Science Education* (pp. 183–202). Cham: Springer International Publishing.

Heath, S. B. (1983). *Ways with Words: Language, Life, and Work in Communities and Classrooms.* Cambridge, England: Cambridge University Press.

Hegarty, M., Carpenter, P. A., & Just, M. A. (1991). Diagrams in the Comprehension of Scientific Texts. In R. Barr, M. L. Kamil, P. B. Mosenthal, & P. D. Pearson (Eds.) *Handbook of Reading Research* (Vol. 2, pp. 641–668). Hillsdale, NJ: Lawrence Erlbaum Associates, Inc.

Hempel, C. G., & Oppenheim, P. (1948). Studies in the logic of explanation. *Philosophy of Science, 15*(2), 135–175.

Henderson, T. (2007). Multimedia Physics Studios: Energy Transformation on a Roller Coaster. Retrieved from www.physicsclassroom.com/mmedia/energy/ce.cfm

Heywood, D. (2002). The place of analogies in science education. *Cambridge Journal of Education, 32*(2), 233–247. doi:10.1080/03057640220147577

Hmelo-Silver, C. E., Chinn, C. A., O'Donnell, A. M., & Chan, C. (2013). *The International Handbook of Collaborative Learning.* New York: Routledge.

Hoban, G., Nielsen, W., & Shepherd, A. (2013). Explaining and communicating science using student-created blended media. *Teaching Science, 59*, 32–35.

Horwood, R. H. (1988). Explanation and description in science teaching. *Science Education*, 72(1), 41–49. doi:10.1002/sce.3730720104

Hubber, P., Tytler, R., & Haslam, F. (2010). Teaching and learning about force with a representational focus: pedagogy and teacher change. *Research in Science Education*, 40(1), 5–28. doi:10.1007/s11165-009-9154-9

Hyland, K. (2015). Metadiscourse. In K. Tracy, C. Ilie, & T. Sandel (Eds.), *The International Encyclopedia of Language and Social Interaction* (pp. 997–1006). Hoboken, NJ: Wiley.

Hyland, K., & Tse, P. (2004). Metadiscourse in academic writing: a reappraisal. *Applied Linguistics*, 25(2), 156–177. doi:10.1093/applin/25.2.156

Jalilifar, A., & Alipour, M. (2007). How explicit instruction makes a difference: metadiscourse markers and EFL learners' reading comprehension skill. *Journal of College Reading and Learning*, 38(1), 35–52. doi:10.1080/10790195.2007.10850203

Jewitt, C. (2008). Multimodality and literacy in school classrooms. *Review of Research in Education*, 32, 241–267.

Jordan, B., & Henderson, A. (1995). Interaction analysis: foundations and practice. *The Journal of the Learning Sciences*, 4(1), 39–103.

Kelly, G. J. (2007). Discourse in Science Classrooms. In S. K. Abell & N. G. Lederman (Eds.), *Handbook of Research on Science Education* (pp. 443–469). Mahwah, NJ: Lawrence Erlbaum Associates.

Kim, M., & Roth, W.-M. (2014). Argumentation as/in/for dialogical relation: a case study from elementary school science. *Pedagogies: An International Journal*, 9(4), 300–321. doi:10.1080/1554480X.2014.955498

Klette, K. (2012). *The Role of Theory in Educational Research*. Oslo, Norway. Retrieved from www.forskningsradet.no/siteassets/publikasjoner/1253979441594.pdf.

Knain, E. (2015). *Scientific Literacy for Participation: A Systemic Functional Approach to Analysis of School Science Discourses*. Rotternam, The Netherlands: Sense Publishers.

Kress, G. (2003). *Literacy in the New Media Age*. London: Routledge.

Kress, G., Jewitt, C., Ogborn, J., & Tsatsarelis, C. (2001). *Multimodal Teaching and Learning: The Rhetorics of the Science Classroom*. London: Continuum.

Kress, G., Jewitt, C., Ogborn, J., & Tsatsarelis, C. (2014). *Multimodal Teaching and Learning: The Rhetorics of the Science Classroom* (2nd ed.). New York: Bloomsbury Academic.

Kress, G., & van Leeuwen, T. (1996). *Reading Images: The Grammar of Visual Design*. London and New York: Routledge.

Kress, G., & van Leeuwen, T. (2006). *Reading Images: The Grammar of Visual Design* (2nd ed.). London and New York: Routledge.

Kristeva, J. (1980). *Desire in Language: A Semiotic Approach to Literature and Art*. New York: Columbia University Press.

Kuhn, T. S. (1962). *The Structure of Scientific Revolutions*. Chicago: University of Chicago Press.

Larsson, J., Airey, J., Danielsson, A. T., & Lundqvist, E. (2018). A fragmented training environment: discourse models in the talk of physics teacher educators. *Research in Science Education*. doi:10.1007/s11165-018-9793-9

Latour, B. (1987). *Science in Action: How to Follow Scientists and Engineers through Society*. Cambridge, MA: Harvard University Press.

Latour, B. (2005). *Reassembling the Social: An Introduction to Actor-Network-Theory*. Oxford and New York: Oxford University Press.

Latour, B., & Woolgar, S. (1979). *Laboratory Life: The Construction of Scientific Facts.* Princeton, NJ: Princeton University Press.

Lave, J., & Wenger, E. (1991). *Situated Learning: Legitimate Peripheral Participation.* Cambridge, England: Cambridge University Press.

Lee, J. J., & Subtirelu, N. C. (2015). Metadiscourse in the classroom: a comparative analysis of EAP lessons and university lectures. *English for Specific Purposes, 37*, 52–62.

Lee, L. W., & Goh, Y. (2017). Use of Premise-Reasoning-Outcome (PRO) in structuring students' answers for open-ended questions. *US-China Education Review B, 7*(4), 195–199. doi:10.17265/2161-6248/2017.04.003

Lemke, J. L. (1984). *Semiotics and Education.* Toronto: Victoria College/Toronto Semiotic Circle Monographs.

Lemke, J. L. (1990). *Talking Science: Language, Learning and Values.* Norwood, NJ: Ablex.

Lemke, J. L. (1992). Intertextuality and educational research. *Linguistics and Education, 4*, 257–267.

Lemke, J. L. (1995). *Textual Politics: Discourse and Social Dynamics.* London and Bristol, PA: Taylor & Francis.

Lemke, J. L. (1998). Multiplying meaning: visual and verbal semiotics in scientific text. In J. Martin & R. Veel (Eds.), *Reading Science* (pp. 87–113). New York: Routledge.

Lemke, J. L. (2000). Multimedia literacy demands of the scientific curriculum. *Linguistics and Education, 10*(3), 247–271.

Lemke, J. L. (2001). Articulating communities: sociocultural perspectives on science education. *Journal of Research in Science Teaching, 38*(3), 296–316.

Lemke, J. L. (2002). Across the scales of time: artifacts, activities, and meanings in ecosocial systems. *Mind, Culture and Activity, 7*(4), 273–290.

Lemke, J. L. (2003). Mathematics in the Middle: Measure, Picture, Gesture, Sign, and Word. In M. Anderson (Ed.), *Educational Perspectives on Mathematics as Semiosis: From Thinking to Interpreting to Knowing* (pp. 215–234). Ottawa: Legas.

Lim, V. F. (2019). Analysing the teachers' use of gestures in the classroom: A systemic functional multimodal discourse analysis approach. *Social Semiotics, 29*(1), 83–111. doi:10.1080/10350330.2017.1412168

Lin, A. M. Y., & He, P. (2017). *Concept + Language Mapping (CLM) as a Strategy in Integrating Language Support into Science Teaching: Translating Lemke's Theory of "Thematic Patterns" into Pedagogical Practice.* Paper presented at the Annual General Meeting of the Hong Kong Association of Applied Linguistics (HAAL), Hong Kong.

Lincoln, Y. S., & Guba, E. G. (1985). *Naturalistic Inquiry.* Newbury Park, CA: Sage.

Lo, Y. Y., & Lin, A. M. Y. (2019). Teaching, learning and scaffolding in CLIL science classrooms. *Journal of Immersion and Content-Based Language Education, 7*(2), 151–165. doi:10.1075/jicb.00006.lo

Lo, Y. Y., Lin, A. M. Y., & Cheung, T. C. L. (2018). Supporting English-as-a-Foreign-Language (EFL) Learners' Science Literacy Development in CLIL: A Genre-Based Approach. In K. S. Tang & K. Danielsson (Eds.), *Global Developments in Literacy Research for Science Education* (pp. 79–95). Cham, Switzerland: Springer.

Lortie, D. C. (1975). *Schoolteacher; a Sociological Study.* Chicago: University of Chicago Press.

Martin, J. R. (2014). Evolving systemic functional linguistics: beyond the clause. *Functional Linguistics, 1*(1), 3. doi:10.1186/2196-419X-1-3

Martin, J. R., & Rose, D. (2007). *Working with Discourse: Meaning beyond the Clause* (2nd ed.). London, England: Continuum.

Martinec, R. (2000). Types of process in action. *Semiotica, 130*(3–4), 243–268.

McNeill, D. (1992). *Hand and Mind: What Gestures Reveal about Thought.* Chicago: University of Chicago Press.

McNeill, K. L., & Krajcik, J. (2008). Scientific explanations: characterizing and evaluating the effects of teachers ' instructional practices on student learning. *Journal of Research in Science Teaching, 45*, 53–78. doi:10.1002/tea

McSharry, G., & Jones, S. (2000). Role-play in science teaching and learning. *School Science Review, 82*, 73–82.

Mehan, H. (1979). *Learning Lessons: Social Organization in the Classroom.* Cambridge, MA: Harvard University Press.

Mercer, N. (2000). *Words and Minds: How We Use Language to Think Together.* London and New York: Routledge.

Mercer, N., Dawes, L., Wegerif, R., & Sams, C. (2004). Reasoning as a scientist: ways of helping children to use language to learn science. *British Educational Research Journal, 30*(3), 359–377.

Moje, E. B. (2007). Developing socially just subject-matter instruction: a review of the literature on disciplinary literacy teaching. *Review of Research in Education, 31*, 1–44.

Moje, E. B., Collazo, T., Carrillo, R., & Marx, R. W. (2001). "Maestro, what is 'quality'?": language, literacy, and discourse in project-based science. *Journal of Research in Science Teaching, 38*(4), 469–496.

Morgan, N., & Saxton, J. (1991). *Teaching, Questioning and Learning.* London: Routledge.

Mortimer, E. F., & Scott, P. (2003). *Meaning Making in Secondary Science Classrooms.* Buckingham, England: Open University Press.

National Research Council. (2012). *A Framework for K-12 Science Education: Practices, Crosscutting Concepts, and Core Ideas.* Washington, DC: The National Academies Press.

National Research Council. (2014). *Literacy for Science: Exploring the Intersection of the Next Generation Science Standards and Common Core for ELA Standards, A Workshop Summary.* Washington, DC: The National Academies Press.

New London Group. (1996). A pedagogy of multiliteracies: designing social futures. *Harvard Educational Review, 66*, 60–92.

NGSS Lead States. (2013). *Next generation science standards: For states, by states.* Washington, DC: The National Academies Press.

Norris, S. P., Phillips, L. M., Smith, M. L., Guilbert, S. M., Stange, D. M., Baker, J. J., & Weber, A. C. (2008). Learning to read scientific text: do elementary school commercial reading programs help? *Science Education, 92*(5), 765–798. doi:10.1002/sce.20266

Novak, J. D. (2010). *Learning, Creating, and Using Knowledge: Concept Maps as Facilitative Tools in Schools and Corporations* (2nd ed.). New York: Routledge.

Novak, J. D., Bob Gowin, D., & Johansen, G. T. (1983). The use of concept mapping and knowledge vee mapping with junior high school science students. *Science Education, 67*(5), 625–645. doi:10.1002/sce.3730670511

O'Connor, M. C., & Michaels, S. (1993). Aligning academic task and participation status through revoicing: analysis of a classroom discourse strategy. *Anthropology & Education Quarterly, 24*(4), 318–335.

O'Halloran, K. L. (2000). Classroom discourse in mathematics: a multisemiotic analysis. *Linguistics and Education, 10*(3), 359–388.

Ochs, E., & Taylor, C. (1992). Science at Dinner. In C. J. Kramsch & S. McConnell-Ginet (Eds.), *Text and Context: Cross-Disciplinary Perspectives on Language Study* (pp. 29–45). Lexington, MA: D.C. Heath.

OECD. (2016). PISA 2015 results in focus. Retrieved from. www.oecd.org/pisa/pisa-2015-results-in-focus.pdf.

Osborne, J. F. (2019). Not "hands on" but "minds on": a response to Furtak and Penuel. *Science Education, 103*(5), 1280–1283. doi:10.1002/sce.21543

Osborne, J. F., Erduran, S., & Simon, S. (2004). Enhancing the quality of argumentation in school science. *Journal of Research in Science Teaching, 41*(10), 994–1020. doi:10.1002/tea.20035

Osborne, J. F., & Patterson, A. (2011). Scientific argument and explanation: a necessary distinction? *Science Education, 95*, 627–638. doi:10.1002/sce.20438

Pappas, C., Varelas, M., Barry, A., & Rife, A. (2004). Dialogic Inquiry around Information Texts: The Role of Intertextuality in Constructing Scientific Understandings in Urban Primary Classrooms. In N. Shuart-Faris & D. Bloome (Eds.), *Uses of Intertextuality in Classroom and Educational Research* (pp. 93–146). Greenwich: Information Age Pub.

Park, J., Chang, J., & Tang, K. S. (in preparation). Plan-Draw-Evaluate (PDE) pattern in students' collaborative drawing: interaction between visual and verbal modes of representation.

Park, J., Chang, J., Tang, K.-S., Treagust, D. F., & Won, M. (2020). Sequential patterns of students' drawing in constructing scientific explanations: focusing on the interplay among three levels of pictorial representation. *International Journal of Science Education*, 1–26. doi:10.1080/09500693.2020.1724351

Peirce, C. S. (1986). *Writings of Charles S. Peirce: A Chronological Edition*. Bloomington: Indiana University Press.

Popper, K. (1963). *Conjectures and Refutations: The Growth of Scientific Knowledge*. New York: Routledge.

Prain, V., & Tytler, R. (2012). Learning through constructing representations in science: A framework of representational construction affordances. *International Journal of Science Education, 34*(17), 2751–2773. doi:10.1080/09500693.2011.626462

Prain, V., & Waldrip, B. (2006). An exploratory study of teachers' and students' use of multi-modal representations of concepts in primary science. *International Journal of Science Education, 28*(15), 1843–1866.

Putra, G. B. S., & Tang, K.-S. (2016). Disciplinary literacy instructions on writing scientific explanations: a case study from a chemistry classroom in an all-girls school. *Chemistry Education Research and Practice, 17*(3), 569–579. doi:10.1039/c6rp00022c

Quillin, K., & Thomas, S. (2015). Drawing-to-learn: a framework for using drawings to promote model-based reasoning in biology. *CBE Life Sciences Education, 14*(1), es2. doi:10.1187/cbe.14-08-0128

Rappa, N. A., & Tang, K.-S. (2018). Integrating disciplinary-specific genre structure in discourse strategies to support disciplinary literacy. *Linguistics and Education, 43*, 1–12. doi:10.1016/j.linged.2017.12.003

Reveles, J. M., Cordova, R., & Kelly, G. J. (2004). Science literacy and academic identity formulation. *Journal of Research in Science Teaching, 41*(10), 1111–1144. doi:10.1002/tea.20041

Roberts, D. (2007). Scientific Literacy/Science Literacy. In S. K. Abell & N. G. Lederman (Eds.), *Handbook of Research on Science Education* (pp. 729–780). Mahwah, N.J: Lawrence Erlbaum.

Roberts, D., & Bybee, R. (2014). Scientific Literacy, Science Literacy, and Science Education. In N. G. Lederman & S. K. Abell (Eds.), *Handbook of Research on Science Education*. (Vol. II, pp. 545–558). New York: Routledge.

Rogowsky, B. A., Calhoun, B. M., & Tallal, P. (2015). Matching learning style to instructional method: effects on comprehension. *Journal of Educational Psychology, 107*(1), 64–78. doi:10.1037/a0037478

Roth, W.-M., & Tobin, K. (1997). Cascades of inscriptions and the re-presentation of nature: how numbers, tables, graphs, and money come to re-present a rolling ball. *International Journal of Science Education, 19*(9), 1075–1091.

Sacks, H., Schegloff, E. A., & Jefferson, G. (1974). A simplest systematics for the organization of turn-taking for conversation. *Language, 50*(4), 696–735.

Sandoval, W. A., & Millwood, K. A. (2005). The quality of students' use of evidence in written scientific explanations. *Cognition and Instruction, 23*, 23–55.

Schiffrin, D. (1980). Meta-talk: organizational and evaluative brackets in discourse. *Sociological Inquiry, 50*(3–4), 199–236.

Scott, P. (1998). Teacher talk and meaning making in science classrooms: A Vygotskian analysis and review. *Studies in Science Education, 32*, 45–80.

Shanahan, M. C. (2012). Reading for Evidence through Hybrid Adapted Primary Literature. In S. P. Norris (Ed.), *Reading for Evidence and Interpreting Visualizations in Mathematics and Science Education* (pp. 41–63). Rotterdam: SensePublishers.

Sinclair, J. M., & Coulthard, M. (1975). *Towards an Analysis of Discourse: The English Used by Teachers and Pupils*. London: Oxford University Press.

Stake, R. E. (2000). Case studies. In N. Denzin & Y. Lincoln (Eds.), *Handbook of qualitative research* (2nd ed., pp. 435–454). London: Sage.

Stoner, M. R. (2007). PowerPoint in a new key. *Communication Education, 56*(3), 354–381.

Strebe, J. D. (2017). *Engaging Students Using Cooperative Learning*. London: Routledge.

Tang, K. S. (2011a). *Hybridizing Cultural Understandings of the Natural World to Foster Critical Science Literacy*. (Doctoral dissertation), University of Michigan, Ann Arbor, MI. Available from ProQuest Dissertations and Theses database. (UMI No. 3476796).

Tang, K. S. (2011b). Reassembling curricular concepts: a multimodal approach to the study of curriculum and instruction. *International Journal of Science and Mathematics Education, 9*, 109–135.

Tang, K. S. (2013a). Instantiation of multimodal semiotic systems in science classroom discourse. *Language Sciences, 37*, 22–35. doi:10.1016/j.langsci.2012.08.003

Tang, K. S. (2013b). Out-of-school media representations of science and technology and their relevance for engineering learning. *Journal of Engineering Education, 102*(1), 51–76. doi:10.1002/jee.20007

Tang, K. S. (2015). The PRO instructional strategy in the construction of scientific explanations. *Teaching Science, 61*(4), 14–21.

Tang, K. S. (2016a). Constructing scientific explanations through premise–reasoning–outcome (PRO): an exploratory study to scaffold students in structuring written explanations. *International Journal of Science Education, 38*(9), 1415–1440. doi:10.1080/09500693.2016.1192309

Tang, K. S. (2016b). The interplay of representations and patterns of classroom discourse in science teaching sequences. *International Journal of Science Education, 38*(13), 2069–2095. doi:10.1080/09500693.2016.1218568

Tang, K. S. (2017). Analyzing teachers' use of metadiscourse: the missing element in classroom discourse analysis. *Science Education, 101*(4), 548–583. doi:10.1002/sce.21275

Tang, K. S. (2019a). *Metalanguage for Describing & Analyzing Multimodal Discourse of Science*. Paper presented at the European Science Education Research Association, Bologna, Italy.

Tang, K. S. (2019b). The role of language in scaffolding content & language integration in CLIL science classrooms. *Journal of Immersion and Content-Based Language Education*, 7(2), 315–328. doi:10.1075/jicb.00007.tan

Tang, K. S. (2019c). Scientific Practices as an Actor-Network of Literacy Events: Forging a Convergence between Disciplinary Literacy and Scientific Practices. In V. Prain & B. Hand (Eds.), *Theorizing the Future of Science Education Research* (pp. 83–98). Cham: Springer International Publishing.

Tang, K. S. (2020). The use of epistemic tools to facilitate epistemic cognition & meta-cognition in developing scientific explanations. *Cognition and Instruction*. doi:10.1080/07370008.2020.1745803

Tang, K. S., & Danielsson, K. (Eds.). (2018). *Global Developments in Literacy Research for Science Education*. Cham, Switzerland: Springer.

Tang, K. S., Delgado, C., & Moje, E. B. (2014). An integrative framework for the analysis of multiple and multimodal representations for meaning-making in science education. *Science Education*, 98(2), 305–326. doi:10.1002/sce.21099

Tang, K. S., Ho, C., & Putra, G. B. S. (2016). Developing Multimodal Communication Competencies: A Case of Disciplinary Literacy Focus in Singapore. In M. Mcdermott & B. Hand (Eds.), *Using Multimodal Representations to Support Learning in the Science Classroom* (pp. 135–158). New York: Springer.

Tang, K. S., & Moje, E. B. (2010). Relating multimodal representations to the literacies of science. *Research in Science Education*, 40(1), 81–85.

Tang, K. S., Tan, A. L., & Mortimer, E. F. (in preparation). Special issue in analysing science classroom discourse. *Research in Science Education*.

Tang, K. S., & Tan, S.-C. (2017). Intertextuality and multimodal meanings in high school physics: written and spoken language in computer-supported collaborative student discourse. *Classroom Discourse*, 8(1), 19–35. doi:10.1080/19463014.2016.1263576

Tang, K. S., Tan, S. C., & Yeo, J. (2011). Students' multimodal construction of work-energy concepts. *International Journal of Science Education*, 33, 1775–1804.

Tang, K. S., Won, M., & Treagust, D. F. (2019). Analytical framework for student-generated drawings. *International Journal of Science Education*. doi:10.1080/09500693.2019.1672906

Thagard, P. (2008). Explanatory Coherence. In E. Adler & L. J. Rips (Eds.), *Reasoning* (pp. 471–513). Cambridge, England: Cambridge University Press.

Tippett, C. D. (2016). What recent research on diagrams suggests about learning with rather than learning from visual representations in science. *International Journal of Science Education*, 38(5), 725–746. doi:10.1080/09500693.2016.1158435

Toulmin, S. E. (1958). *The Uses of Argument*. Cambridge, England: Cambridge University Press.

Treagust, D. F., Duit, R., Joslin, P., & Lindauer, I. (1992). Science teachers' use of analogies: observations from classroom practice. *International Journal of Science Education*, 14(4), 413–422. doi:10.1080/0950069920140404

Unsworth, L. (2001a). Evaluating the language of different types of explanations in junior high school science texts. *International Journal of Science Education*, 23(6), 585–609.

Unsworth, L. (2001b). *Teaching Multiliteracies across the Curriculum: Changing Contexts of Text and Image in Classroom Practice*. Buckingham, England: Open University Press.

van Leeuwen, T. (1998). Music and ideology: notes toward a sociosemiotics of mass media music. *Popular Music and Society*, 22(4), 25.

Van Rooy, W. S., & Chan, E. (2017). Multimodal representations in senior biology assessments: a case study of NSW Australia. *International Journal of Science and Mathematics Education, 15*(7), 1237–1256. doi:10.1007/s10763-016-9741-y

Van Zee, E., & Minstrell, J. (1997). Using questioning to guide student thinking. *The Journal of the Learning Sciences, 6*(2), 227–269.

Vande Kopple, W. J. (2012). The importance of studying metadiscourse. *Applied Research on English Language, 1*(2), 37–44.

Vygotsky, L. (1986). *Thought and Language* (Translation newly rev. and edited/by Alex Kozulin ed.). Cambridge, MA: MIT Press.

Vygotsky, L. (1987). Thinking and Speech. In R. W. Rieber & A. S. Carton (Eds.), *Collected Works of L. S. Vygotsky* (pp.243–285). New York: Plenum.

Wang, C.-Y. (2014). Scaffolding middle school students' construction of scientific explanations: comparing a cognitive versus a metacognitive evaluation approach. *International Journal of Science Education, 37,* 237–271. doi:10.1080/09500693.2014.979378

White, R., & Gunstone, R. F. (1992). *Probing Understanding.* London; New York: Falmer.

Whitebread, D., Coltman, P., Pasternak, D. P., Sangster, C., Grau, V., Bingham, S., . . . Demetriou, D. (2009). The development of two observational tools for assessing metacognition and self-regulated learning in young children. *Metacognition and Learning, 4*(1), 63–85. doi:10.1007/s11409-008-9033-1

Williams, J. M. (1981). *Style: Ten Lessons in Clarity and Grace.* Glenview, IL: Scott Foresman.

Williams, M., & Tang, K. S. (2020). The implications of the non-linguistic modes of meaning for language learners in science: a review. *International Journal of Science Education.* doi:10.1080/09500693.2020.1748249

Williams, M., Tang, K.-S., & Won, M. (2019). ELL's science meaning making in multimodal inquiry: a case-study in a Hong Kong bilingual school. *Asia-Pacific Science Education, 5*(1), 3. doi:10.1186/s41029-019-0031-1

Wragg, E., & Brown, G. (2001). *Questioning in the Secondary School.* London: Routledge.

Wu, S. J., Mensah, F. M., & Tang, K. S. (2018). The Content-Language Tension for English Language Learners in Two Secondary Science Classrooms. In K. S. Tang & K. Danielsson (Eds.), *Global Developments in Literacy Research for Science Education* (pp. 113–130). Cham, Switzerland: Springer.

Yore, L. D. (2018). Commentary on the Expanding Development of Literacy Research in Science Education. In K. S. Tang & K. Danielsson (Eds.), *Global Developments in Literacy Research for Science Education* (pp. 379–397). Cham, Switzerland: Springer.

APPENDIX

Research Methods behind this Book

Research Context

The excerpts of classroom conversation in this book were taken from the following research projects I have led and participated in Singapore and the United States over the years:

S/N	Project Name	Subject	Level	Year	Country
1	*Literacy, Representation, and Disciplinary Knowledge: An Exploratory Multimodal Study of Science Classroom Discourse*	Nanoscience	Grade 6–7	2008	U.S.A.
2	*Forging a Third Space of Everyday Literacies & Multimodal Science*	Physics	Grade 9–12	2009–2010	U.S.A.
3	*Developing Disciplinary Literacy Pedagogy in the Sciences*	Physics, Chemistry	Grade 9–10	2013–2015	Singapore
4	*Formalizing Disciplinary Literacy Teaching in Primary Science through a Writing-to-Learn Approach*	General Science	Grade 4–5	2014–2016	Singapore

These research projects had different objectives and involved teachers and students from various science subjects and levels, ranging from general science in primary schools to physics and chemistry in upper secondary schools. Despite these variations, all these projects shared three things in common. First, they shared a similar focus on the use of language and literacy (including non-verbal representations) to support science teaching and learning. Pedagogically, the teachers understood the importance of language and worked with the researchers to design and enact strategies that could harness language more effectively to enhance student learning. Second, the teachers who took part in the projects were mostly experienced teachers who either volunteered or were nominated by the school leaders to take part in the research. All the participating classes were considered mainstream and typical of most classes in their countries. The students involved were also considered average to above-average ability according to the respective state standards. There were only a few students who just started learning English at the time of the research.

The third commonality was the underlying methodology behind the collection and analysis of data in these research projects, which I will elaborate further.

Research Design and Data Collection

The research methodology or logic of inquiry that drove the data collection and analysis in all the research projects can be characterized as ethnographic, or more specifically microethnography (Erickson, 2006). Ethnographic method involves an extended period of observing a group of people in their community to study an aspect of their shared cultural practices they developed over time. The community in my case was obviously the classroom, and the cultural aspect I was studying was the participants' use of language in the context of science; not just in terms of their abilities in using language, but also the larger discourse in the form of behavioral norms, social practices, and values that come with learning and using a language. The research was microethnographic because it involved a close study of the participants' interaction through audiovisual recordings at a micro-analytical timescale (Garcez, 2017). By studying how the interaction unfolded moment by moment, the goal was to understand how different discourse practices were enacted in the classrooms, and subsequently documented them for others to reproduce an exemplary practice or change an existing practice (Erickson, 2006). This goal from microethnography aligned well with the research-practice nexus objective of this book.

Video recordings of rich classroom interactions thus provided the primary data source in all the projects. This involved going into the classrooms to observe and record classroom events from the beginning to the end of a lesson, and also over a continuous series of lessons bounded by a common science topic or curricular program. This continuity over a topic or program usually lasted more than five lessons for a single topic and months for several topics in an academic semester or year. It was important to maintain a continuity in order to provide a rich context for interpreting and making judgment over what was going on in many instances, such as a previous content development that shaped an ongoing discussion or a teacher's decision in an instructional activity to prepare a future lesson.

Two video cameras were often used to record the lessons. One camera was mounted on a tripod at the back of the classroom and focused on the teacher.

The second camera, focusing on student interaction, was either mounted on a tripod and fixed on a preselected group of students or carried by the observer to record any interesting interactions as they occurred. For the projects in the United States, I was present in all the classroom observations. For the projects in Singapore, due to the scale involved, the observation and recording were distributed among three research assistants and myself. Besides video recordings as the primary data source, secondary data sources such as field notes, students' artifacts, and student and teacher interviews were also collected as part of the ethnographic data collection.

Data Analysis

The work of discourse analysis began once video data of the classroom interaction had been collected. Discourse analysis is not a singular technique that is universally shared and followed among researchers, unlike commonly known techniques such as constant comparative method or factor analysis. It is more accurate to describe discourse analysis as an umbrella term for a variety of approaches from different intellectual traditions developed to analyze language-in-use and talk-in-interaction. Within science education, the common traditions include conversation analysis, critical discourse analysis, discursive psychology, ethnography of communication, multimodal discourse analysis, practical epistemology analysis, symbolic interactionism, and systemic functional linguistics (Tang, Tan, & Mortimer, in preparation).

Despite the variety of discourse analysis approaches, there are generally two levels of analysis that most classroom-based researchers have to deal with, in what I have called macro and micro timescale analysis in my projects (Tang, 2016b; Tang, Delgado, & Moje, 2014). Although these two levels of discourse analysis require different treatment on the data, they should not be regarded as separate analyses with independent purposes, but as two mutually dependent aspects of a coherent analysis. The macro timescale event analysis generates and provides the situated context that is essential for interpretation of meaning-making at a micro timescale. Conversely, the purpose of analyzing the micro-discursive details of several meaning-making instances allows us to make sense of how the events at the macro timescale unfolded and connected to one another.

Macro-event Analysis

The first procedural step in a macro timescale event analysis was to view the entire videos and divide the continuous data stream into discrete segments of meaningful interaction called episodes. Each episode was determined by clear boundaries demarcating prominent shifts that occurred in the interaction, such as a discernible change in the participation structure or the nature of the talk or activity involved (Erickson, 1992). Across the projects, the average duration of an episode was three to five minutes. Each episode was then coded and tagged to different categories of analytical interests. The categories in my projects usually involved literacy event (e.g., teacher talk, group talk, student writing), dominant representation (e.g., worksheet, model, presentation slide), and science content. A brief description of the event and anything of analytical interest was also tagged in each episode. A video qualitative analysis software called *Transana* was used to facilitate the viewing, segmentation, coding, and transcription process.

After the segmentation, a content listing (Jordan & Henderson, 1995) indexing the categories and description of every episode with their respective timestamp in the video was then generated. This content listing served to provide an overall contextual basis for subsequent analysis as well as facilitate a purposeful sampling and selection of interesting cases (Stake, 2000) for further analysis. Due to the amount of video data collected, not every episode was and could be transcribed. Thus, the selection of interesting cases was essential in making strategic decisions over what to transcribe.

In general, a combination of two factors shaped the choice of what to select for further analysis. The first factor was informed by the research questions and key theoretical constructs from the projects to systematically look for relevant episodes to analyze. The second factor was the identification of "hot spots" that were out of the ordinary in most classroom interactions, for instance an ideational conflict between a teacher and student or an intensive group discussion that demonstrated deep conceptual learning. Our identification of what is interesting in a hot spot is usually guided by our research questions and constructs, so this tends to reinforce the selection from the first factor. But sometimes it can be productive to look out for hot spots outside our research plan and objective. For instance, although metacognition was never a focus in my research projects, several interesting hot spots involving a particular teacher's use of metacognition led to a productive study of the connection between epistemic language and metacognition (see Tang, 2020).

Micro-discursive Analysis

After numerous episodes of analytical interest had been identified and transcribed, the next phase of micro-level discourse analysis began. This involved a line-by-line analysis of the participants' utterances and actions, as demonstrated in the 37 excerpts presented in this book. By default, what is counted as the "unit of analysis" for a line is usually a single turn as determined from the turn-taking change of speakers. This was the case in Chapter 3 when we were analyzing interaction pattern. However, this was not the case for other analytical purposes. Other units smaller than a turn were required when a turn (particularly a long one from the teacher) could contain several discursive purposes. For instance, in Chapter 5, a turn could serve several metadiscourse functions. And in Chapters 4 and 8, when we were analyzing ideational content, the unit of analysis for a line would be a grammatical clause. This is because, according to Halliday's (1985) SFL, a clause functions as the basic semantic unit that constructs our sense of experience.

On a transcript with every line demarcated, the next step was to examine and code all the lines in the episode. Various coding schemes were used depending on the specific purposes. For the purpose of identifying the participants' interactions in Chapter 3, I used a coding scheme based largely on Chin's (2006) study to examine discursive moves (e.g., initiate, response, evaluate, follow-up). For the purpose of identifying thematic content in Chapter 4, I used a coding scheme (Table 4.1) that was adapted from Lemke (1990) to map the semantic relationship made in every few lines of speech. In Chapter 5 focusing on metadiscourse, I used a coding framework (summarized in Table 5.1) developed in a previous study (Tang, 2017). In addition, the study in Tang (2017) also demonstrated how the three coding schemes – discursive, semantic, metadiscourse – were combined in the discourse analysis of several episodes, including Excerpts 4.2 and 5.7 which were reproduced in this book.

For Chapter 6, the focus on genre and metalanguage emerged from various studies on the development, enactment, and application of PRO (premise-reasoning-outcome) in classroom discourse, as reported in Tang (2015), Tang (2016a), Putra and Tang (2016), and Rappa and Tang (2018). Specifically, in terms of the application of discourse analysis on genre and metalanguage, the coding scheme used to analyze the data was drawn from and described in Rappa and Tang (2018). Thus, more information on how the excerpts in that chapter, particularly Excerpt 6.3, were coded and analyzed can be found in Rappa and Tang (2018).

As for the focus on representations in Chapters 7 and 8, the analyses were drawn from multiple studies and projects. Chapter 7 examines the larger timescale of translating representations; thus it is mostly based on a macro-event analysis instead of a micro-discursive analysis. The study in Tang (2016b) provides more methodological details over how the translation pattern (from concrete to pictorial to abstract, and from dialogic to authoritative) was analyzed and documented in two classrooms. Chapter 8 examines the micro timescale of integrating representations. For this purpose, in conjunction with the linguistic framework presented in Chapter 4, I also used a visual framework (Table 8.1) developed in Tang, Won, and Treagust (2019) to analyze visual diagrams. Many of the visual examples were drawn from that study. As for the analysis of multimodal thematic patterns, more methodological details can be found in Tang (2011b) and Tang, Tan, and Yeo (2011). In particular, Excerpt 8.1 was reproduced from Tang, Delgado, and Moje (2014) and re-analyzed to demonstrate the use of the visual analytical framework in multimodal analysis.

Finally, the purpose of coding every line in an episode was not to generate a frequency count of or provide an isolated evidence of whatever was being studied. Instead, the purpose of the coding was to examine patterns of discourse that were manifested in the classroom data. As previously mentioned, this is essentially the key to discourse analysis as a methodological approach in classroom discourse – to identify social patterns in the use of language that shape and are shaped by the way we think, act, and make meanings. Methodologically, any pattern from one episode is obviously not sufficient to generalize into a social pattern. This is why many episodes selected and re-selected from the previous macro-event analysis are needed in order to look for recurring patterns, until a saturation point is reached where no further understanding or new insight can be derived from reviewing the video data (Lincoln & Guba, 1985). It was primarily through this analytical approach that the patterns presented in this book were studied and revealed, specifically the interaction, thematic, narrative, genre, and multimodal patterns from Chapters 3 to 8.

INDEX